Vinícius Pereira dos Santos
Dirceu Pratissoli

Action of Insecticides and Fungicides Used on Tomato Crops

Vinícius Pereira dos Santos
Dirceu Pratissoli

Action of Insecticides and Fungicides Used on Tomato Crops

Trichogramma pretiosum Riley, 1879 (Hymenoptera: Trichogrammatidae)

ScienciaScripts

Imprint

Cover image: www.ingimage.com

This book is a translation from the original published under ISBN 978-3-330-75931-2.

Publisher:
Sciencia Scripts
is a trademark of
Dodo Books Indian Ocean Ltd. and OmniScriptum S.R.L publishing group

120 High Road, East Finchley, London, N2 9ED, United Kingdom
Str. Armeneasca 28/1, office 1, Chisinau MD-2012, Republic of Moldova, Europe
Printed at: see last page
ISBN: 978-620-8-30691-5

SUMMARY

I dedicate this work to my family in Alegre and Sao Mateus, especially my beloved father (*in memoriam*). Thank you to everyone who has always supported me in achieving my goals.

ACKNOWLEDGMENTS

I thank our Lord God for another victory in my life.

To my family for their support at every moment of my life. In particular, to my father Benedito José dos Santos, who is no longer here, but who always helped me and believed in my dreams. To my mother Geilza Pereira dos Santos for being the great love of my life.

To my beloved wife and partner Érika Aparecida Silva de Freitas for her understanding, friendship, companionship and encouragement. To my daughter for the mess and the walks in the squares of this city of Alegre. To my dear "son" and friend Hugo Tristao de Freitas Sartore for his struggle and dedication to his studies. To my favorite sister-in-law Cristina Freitas Balarini and her family. To my Alegre parents Rosemburgo Càssia de Freitas and Wilma Alves da Silva.

To Prof. Dr. Dirceu Pratissoli for his guidance, teaching and the opportunity to study for a Master's degree in Plant Production at UFES.

The Board of Examiners: Prof. Dr. Dirceu Pratissoli, Prof. Dr. Ulysses Rodrigues Vianna, Dr. José Salazar Zanùncio Junior (INCAPER), Dr. Ramon Santos de Minas and Dr. Hugo Bolsoni Zago.

To my friends at NUDEMAFI, starting with Tia Carlota and Leonardo Mardgan, Alegre's greatest piper and fisherman. To the undergraduate students, Lorena, Priscila, Ingrid, Amanda, Wilson, Higor, Débora Fornazier and Leam. To my postgraduate friends Carlos Magno, Eduardo Grecco, Fernando Zinger, Flàvio Neves, Joao Paulo, José Romàrio de Carvalho, Kharen Priscila, Luziani Rezende, Vando Miossi and Ramon Santos de Minas.

To Professor Dr. Rosembergue Nipes Bragança.

To my lifelong friends and the friends I made in Alegre during my master's degree: Ana Clàudia Nascimento (PhD), Arieli Altoé (Master's degree), Lucas Contarato Pilon (Master's degree), Lauana Pelanda (Master's degree), Carlos Magno Ramos (PhD), Débora Ferreira Melo (PhD), Flàvio Neves Celestino (PhD), Eduardo Domingos Greco (PhD), Thiago Kloss (PhD student) and Kharen Priscila de Oliveira Silva Salomao (Master's degree).

To my great friend and brother Rodrigo da Silva from Atual Sistemas in Alegre.

For the Coordination of the CCAUFES Plant Production Program - Prof. Dr. Edvaldo Fialho dos Reis.

The Secretary of the CCAUFES Plant Production Program - Madalena Caetano Capucho.

To old friends from EAF-COL, now IFES Campus Itapina.

To the Financier of Studies and Projects (FINEP), the National Council for Scientific and Technological Development (CNPq) for financing the project and the Coordination for the Improvement of Higher Education Personnel (CAPES) for the master's scholarship.

To the old and new employees of ESAES, CAUFES and CCAUFES.

To all those who make education the foundation of a better and fairer future.

AUTHOR'S BIOGRAPHY

Vinicius Pereira dos Santos was born in Sao Mateus, state of Espirito Santo, on May 19, 1982, to Benedito José dos Santos (*in memoriam*) and Geilza Pereira dos Santos. He studied at the Escola Estadual de 1° e 2° Grau "Polivalente" until the 8th grade and at the Escola Agrotécnica Federal de Colatina EAF-COL (now IFES Campus Itapina) where he completed high school and vocational training as an Agricultural Technician between 1998 and 2000. At the age of 18, he entered the Agricultural Engineering course at the Federal University of Espirito Santo at the Alegre Campus, now the Agricultural Sciences Center of the Federal University of Espirito Santo CCA-UFES. During his undergraduate studies, he was a computer science monitor from the second to the tenth period. At the age of 26, he obtained a degree in Agricultural Engineering. He then enrolled as a special student on the Postgraduate Course in Plant Production in 2009, in the city of Alegre-ES. On July 17, 2010, his first child, Leticia Freitas dos Santos, was born. In March 2011, he joined the Postgraduate Program in Plant Production as a regular student, under the noble guidance of Prof. Dr. Dirceu Pratissoli. On February 28, 2013, he defended his dissertation to obtain a Master's degree in Plant Production, Area of Concentration in Plant Health (Entomology).

"You don't measure a man's worth by his clothes or his possessions, the true worth of a man is his character, his ideas and the nobility of his ideals."

Charles Chaplin

CO-AUTHOR'S BIOGRAPHY

He holds a degree in Agronomy from the Federal University of Espirito Santo (1981), a master's degree in Agronomy (Plant Science) from the Federal University of Lavras (1985) and a doctorate in Entomology from the University of Sao Paulo (1995). He is currently a Class E Full Professor at the Federal University of Espirito Santo; a researcher at the National Council for Scientific and Technological Development; a member of the agricultural health defense commission. - Federal Superintendence of Agriculture, Livestock and Supply in ES; member of the technical committee for the integrated production of table tomatoes - MAPA. He has experience in Agronomy, with an emphasis on Agricultural Entomology, working mainly on the following subjects: biological control with *Trichogramma* and phytosanitary pest management.

GENERAL SUMMARY

SANTOS, Vinicius Pereira dos, M Sc. Federal University of Espírito Santo, February 2013. **Action of insecticides and fungicides used on tomato crops on** *Trichogramma pretiosum* Riley, 1879 (Hymenoptera: Trichogrammatidae)**.**

Advisor: Prof. Dr. Dirceu Pratissoli. **Co-Orientator:** Prof. Dr. Hugo Bolsoni Zago

The aim of this study was to evaluate the effect of the main pesticides registered for tomato cultivation on the parasitism of the egg parasitoid *Trichogramma pretiosum*. 29 fungicides and 30 insecticides were evaluated for their selectivity. The insecticides and fungicides were separated into toxicological classes according to the reduction in parasitism capacity, according to the scale proposed by members of the IOBC/WPRS. The results showed that the insecticides Tiger® 100g/L (PIRIPROXIFEM), Nomolt® 150g/L (TEFLUBENZURON), Mospilan® 200g/Kg (ACETAMIPRID), Karate Zeon® 50g/L (LAMBDA-CIALOTRINA) and Ampligo® 100g/L (CHLORANTRANILIPROLE) + 50g/L (LAMBDA-CIALOTRINA) were considered innocuous (Class 1) for the parasitoid *T. pretiosum. pretiosum*. For the fungicides analyzed, it was concluded that Kocide WDG® 538g/Kg (COBRE HYDROXIDE), Nativo® 200g/L (TEBUCONAZOLE) + 100g/L (TRIFLOXYSTROBIN), Censor® 500g/L (FENAMIDONE), Manzate® 800g/Kg (MANCOZEB), Orthocide 500® 500g/Kg (CAPTAN), Amistar 500 WG® 500g/Kg (AZOXYSTROBIN), Frowncide 500 SC® 500g/L (FLUAZINAM), Fegatex® (BENZALCONIUM CHLORIDE), Redshield 750® 860g/Kg (CUPROSOXIDE), Proplant® 722g/L (PROPAMOCARB CHLORIDRATE), Isatalonil 500 SC® (CHLOROTALONIL), Recop® 840g/Kg (COVER OXYCLORIDE), Bion 500 WG® 500g/Kg (ACIBENZOLAR-S-METHYL) and Curzate Br® 80g/Kg (CYMOXANIL) + 640g/Kg (MANCOZEB) were found to be harmless fungicides (Class 1) to the parasitoid *T. pretiosum. pretiosum*. After analyzing the data, it was possible to establish a list of insecticides and fungicides with potential for use in the integrated management of tomato pests by making the use of chemical control compatible.

Keywords: Selectivity. Susceptibility. Biological control.

CHAPTER I

1.1 GENERAL INTRODUCTION

In 2011, the world tomato crop totaled 10.8 million tons, with a cultivated area of 4.62 million hectares and an average yield of 27.3 t/ha (FAOSTAT, 2012). The world's largest tomato producer is China (44.9%), followed by India (15.55%), the United States (11.66%), Turkey (10.27%) and Egypt (7.5%). China also has the largest cultivated area (31.38%), followed by India (10.36%), Turkey (5.80%), Egypt (4.20%) and the USA (3.79%). In 2011, Brazil ranked eighth in production (0.44%), 14th in cultivated area (2.67%) and sixth in productivity, with an average yield of 62.87 tons/ha, 117.53% higher than the world average (FAOSTAT, 2012). Between 1983-85 and 2003-05, world *per capita* tomato consumption grew by 36%, from 14 kg to 19 kg per inhabitant per year, according to data from the United Nations Food and Agriculture Organization (FAOSTAT, 2012).

In Brazil, according to AGRIANUAL (2012), the states of Goiás, São Paulo and Minas Gerais are the largest tomato producers in the country, with 52% of the total area cultivated. Among the regions producing this vegetable, the southeast region stands out from the rest as it accounts for approximately 35.62% of all Brazilian production. In 2008, the state of Goiás produced over 83.74 tons/ha, while Espirito Santo produced only 68.25 tons/ha. Data from 2011 shows some growth for Espirito Santo with average productivity of 70.22 tons/ha, an increase of 2.88% in average productivity.

According to AGRIANUAL (2012), the 2011 national table tomato crop was estimated at 4,103,435 tons harvested from an area of 65,262 ha, with an average yield of 62.87 tons/ha, according to the systematic survey of agricultural production. The state of Espirito Santo had an estimated production of 133,987 tons in an area of 1,908 ha, with an average productivity of 70.22 tons/ha, above the national average.

The tomato plant (*Lycopersicum esculentum*, Mill.), which is grown in all Brazilian regions, is a crop that stands out for its high investment in inputs and high risk, due to the high infestation of pests and incidence of diseases, both in crops destined for fresh consumption and for industry (SOUZA; REIS, 2003; LUZ et al., 2007). In order to achieve viable production, farmers use pesticides indiscriminately (CARVALHO et al., 2003). Therefore, in general, both agriculture and tomato farming should seek to use methods that aim to be rational and efficient in order to stimulate biological processes and recover the potential of environmental biodiversity (PROTAS, 2003).

The system implemented by man in recent decades has been characterized by the abusive use of agrochemicals which, in addition to contaminating the environment and man, has led to the induction of pest resistance to certain active ingredients. It is of great importance to study and improve pest control methodologies that aim to integrate techniques, making the use of chemical control compatible with biological control (CARMO et al., 2010).

Chemical control is the most widely used method in conventional tomato crop management, but although effective in most cases, pesticides can negatively affect the community of natural enemies, especially parasitoids. In addition, the use of biological control agents in agricultural crops in Brazil is on the increase, with *T. pretiosum* having already been recorded in more than 3,000 hectares of corn in 2008 in the Santa Maria region of Rio Grande do Sul (NAVA; NACHTIGAL, 2010). In addition to mass use, there are some studies with these parasitoids aimed at the biological control of pests of sugar cane (BOTELHO et al., 1999), cotton (BLEICHER; PARRA, 1989), tomatoes (PRATISSOLI; PARRA, 2000), corn (SA; PARRA, 1994), stored grain pests (INOUE; PARRA, 1998) and citrus (MATOS, 2007).

The conventional tomato production system is primarily concerned with quantity, leaving the health quality of this food in the background. Nowadays, there is a great concern about consuming products that are free of or have little pesticide residue. There is therefore a need for the effective and intensive adoption of alternative methods to the excessive use of pesticides. Due to environmental contamination and pesticide residues in food, we can also estimate that the populations living near cultivation areas and urban dwellers are also exposed to the harmful effects of these chemical agents (CARVALHO et al, 2005).

Parasitoids of the genus Trichogramma have been widely used in biological pest control programs (ALMEIDA et al., 2010). Among the various species, *T. pretiosum* stands out due to its wide distribution, being the most commercially exploited parasitoid and its aggressiveness, being reported on 18 hosts and 13 crops (ZUCCHI; MONTEIRO, 1997; PRATISSOLI *et al.*, 2004; DELPUECH *et al.*, 2010).

Based on the holistic view of Phytosanitary Pest Management (PPM), it is of great importance to integrate two or more methods of controlling insect pests in order to achieve better control and a lower impact on the environment. The aim is to integrate methods such as the use of pesticides (insecticides and fungicides) and biological control. The integration of control techniques with more than one tool to control pests and diseases makes it possible to produce to meet demand without compromising the quality of the end product.

Biological and chemical controls need to be made compatible in order to preserve the community of beneficial insects in the production system. In this sense, the working group of the International Organization for Biological Control of Noxious Animals and Plants (IOBC), West Palaearctic Regional Section (WPRS) has established guidelines for conducting bioassays and categorizing pesticides in terms of selectivity, with the following stages: laboratory (adult and immature phases); laboratory/vegetation house (persistence) and field (adult phase) (HASSAN et al., 2000; HASSAN; ABDELGADER, 2001).

According to Goulart et al. (2008), for good results in Integrated Pest Management (IPM) programs, it is important that phytosanitary products are efficient against pest species and selective to non-target insects. Work carried out by Gravena (1992) showed that the selectivity of pesticides is mandatory in the phytosanitary management of pests if biological control is to be fully successful. According to Carvalho et al. (2001), the combination of chemical and biological methods is important for controlling insect pests, allowing for a reduction in the number of applications of phytosanitary products, with greater economy and less environmental impact. Therefore, one of the strategies commonly used in Phytosanitary Pest Management programs is the joint use of biological control agents and selective agrochemicals, since currently the complex of pests attacking a crop cannot be controlled by the biological method alone (RUBERSON; TILLMAN 1999, MEDINA et al., 2003).

In Brazil, however, research into the persistence of the action of pesticides on *T. pretiosum*, using the IOBC/WPRS methodology, has only been carried out for a few fruit trees, in which products used for peach (GIOLO et al., 2008) and apple (NORNBERG, 2009) crops have been tested.

In this study, laboratory bioassays were carried out using the biological control agent *T. pretiosum* on a list of 29 fungicides and 30 insecticides registered for the control of pests and diseases in tomato cultivation, with the aim of establishing a list of insecticides and fungicides with potential for use in biological control.

1.2 LITERATURE REVIEW

1.2.1 Tomato - *Lycopersicum esculentum* Miller (1754)

The tomato is the most industrialized vegetable, employing large amounts of inputs and labour, with a derivatives market that mainly exploits the production of extracts, ready-made sauces and ketchup (KROSS et al., 2001). The tomato crop has accumulated many genetic gains with plant breeding aimed at greater production, post-harvest longevity of the fruit and the possibility of cultivation in more remote regions of the world, but it continues to present many problems, mainly phytosanitary.

It is worth noting that this is a high-value crop that requires constant care, as it is subject to attack by a large number of pests, diseases and physiological disorders. Long-life hybrid seeds are very expensive, costing an average of more than 20,000 reais per 500-gram packet, which can be used to plant 150,000 tomato plants. As an example of this trend, 500 grams of Itapua 800 SAIS double hybrid long-life tomatoes costs 21,000 reais (ISLA, 2013). It is important to note that pesticides account for more than 35% of the total cost of staking tomato crops (MELO 2003).

In the last three decades there has been an increase in production, with a greater contribution from the productivity of industrial tomatoes due to the use of more productive cultivars and hybrids, while the production of tomatoes for *fresh* consumption has only kept pace with population growth (FAOSTAT, 2012).

Data from INCAPER (Instituto Capixaba de Pesquisa, Assistência Técnica e Extensao Rural) from 2010 show that the state of Espirito Santo is self-sufficient in the production of most of the vegetables consumed in the state, with exportable surpluses, tomatoes being the most important vegetable in the state in economic and social terms. The crop is grown all over the state, in an area of approximately 2,000 hectares, involving 700 farms, which produce around 140,000 tons of the fruit a year, generating 10,000 direct jobs and an amount of approximately R$100 million. The social importance of tomatoes lies in the fact that the activity is highly labor-intensive, with an average of five rural workers working each hectare (SEAG- ES, 2012). The technological level of the crop is considered to be very high, as it is on a par with other important producing states such as Goiás, São Paulo, Minas Gerais and Rio de Janeiro. Of the volume of produce that stays in the state, between 20% and 30% passes through the Espirito Santo Supply Center (Ceasa-ES) and, of these, 72% is sold in Greater Vitória, 76% in supermarket chains and kiosks, and 24% in street markets and industrial kitchens. The main producing municipalities are Santa Teresa, Laranja da Terra and Venda nova do Imigrante (INCAPER, 2010).

The high damage potential of tomato pests such as borers and moths has caused concern among tomato growers in recent years, who, in the absence of alternative control methods, use insecticides without technical knowledge and apply them incorrectly. In this context, pests can develop resistance to insecticides, natural enemies have their population reduced or eliminated and toxic residues have caused serious problems for public health and the environment (TAMISO, 2005). Another unfavorable factor is related to the planting of crops in concentrated areas, with the aim of year-round production, which means that pest populations are always high (MELO, 2003).

Currently, agricultural activity requires greater efficiency as it is part of a highly competitive environment, with the need to improve planning, management and production techniques in order to guarantee sustainability. Tomato production is also part of this panorama, as it is a crop that requires constantly evolving technology and specific knowledge for cultivation. For this reason, the less skilled farmer will find it difficult to produce in quantity and quality (BOREL et al., 2010).

1.2.2 Phytosanitary pest management

The concept of Integrated Pest Management (IPM) encompasses the harmonious use of multiple techniques to keep pests below the level of economic damage and to conserve the environment (CORSO et al., 1999). These systems are represented by biological methods, plant resistance, cultural and physical methods, including chemical methods, provided they are used in an appropriate and rational manner (PARRA, 1993).

Integrated pest management (IPM) emerged in the mid-20th century as a response to the harm caused by the improper and unguided use of pesticides. IPM can therefore be defined as the joint use of management techniques in isolation or grouped together to control phytosanitary disorders, analyzing the cost/benefit ratio in line with the interests and/or impact on producers, society and the environment (GALLO et al. 2002, PICANÇO et al. 2004, VALE et al. 2004).

The discovery and synthesis of molecules with insecticidal action during World War II triggered an extraordinary and unbridled development of the chemical industries, resulting in the emergence of countless chemical products for different purposes. The pest control methods used up until then (cultural, biological, physical) were quickly replaced because the chemical method was so fast-acting and efficient in the short term. However, several adverse effects of the indiscriminate use of broad-spectrum pesticides soon emerged, such as the resurgence of primary pests due to their direct effects on populations and natural enemies, the elevation to primary pest status of pests that were previously of secondary importance, resistance to insecticides (as early as 1946, the first case of resistance to DDT was reported), residues in food, environmental problems arising from the persistence of products in soil and water, bioaccumulation, among others (BOARETTO; BRANDÃO, 2000).

After IPM, new definitions of pest management emerged, and among these: ecological pest management (EPM) demonstrates a singular concern with the interactions that occur in the agro-ecosystem, being defined as: "A *system of actions against insects and mites harmful to agriculture, with emphasis on the preservation and increase of natural enemies and the use of environmental management techniques, with monitoring and sampling as support for decision-making*" (GRAVENA; BENVENGA, 2003).

The foundations of both Integrated Control and Integrated Pest Management are based on four elements: the exploration of natural control, the tolerance levels of plants to the damage caused by pests, the monitoring of populations for decision-making and the biology and ecology of the crop and its pests. In Brazil, Integrated Pest Management programs have been implemented for some economically important crops, such as soybeans, cotton, citrus fruits, among other fruit trees, with promising results that imply a reduction in the number of applications, resulting in savings for the farmer and minimizing adverse effects on the environment.

Recently, the Center for Agricultural Sciences at UFES, in view of the need to change the agricultural panorama in the state of Espirito Santo and, above all, to provide rural producers and professionals linked to agriculture with a technological package aimed at using pest management techniques efficiently and in line with current principles of sustainability, founded the Scientific and Technological Development Center for Phytosanitary Pest Management (NUDEMAFI). This center advocates the use of alternative methods to control pests, based on planning the crop before it is planted and evaluating the impact of phytosanitary measures on the agro-ecosystem.

1.2.3 Selectivity of pesticides to natural enemies

Plant health is one of the foundations for agricultural productivity and competitiveness (MOLINA, 2005). Thus, the basic goal of Phytosanitary Pest Management (PPM) is to reduce the use of agrochemicals, making it possible to preserve populations of natural enemies and reduce the risk of contamination, whether human or of the agroecosystem (PRATISSOLI et al. 2007). Tomato crops without the application of pesticides have a rich fauna of predatory and parasitoid insects (GRAVENA, 1991), and the preservation of these species is fundamental for the MFP, due to their high control potential, without increasing the cost of production. The use of biological control agents is an extremely important tool for the development of a Phytosanitary Pest Management Program (PRATISSOLI et al. 2007).

Biological control for these species can be carried out using micro-hymenoptera of the genus Trichogramma. For the tomato moth, the use of *T. pretiosum* is a viable strategy due to the fact that they naturally parasitize this pest species in the mountainous region of the state of Espirito Santo (PRATISSOLI et al., 2003). According to Alvarenga *et al.* (2004), the use of *T. pretiosum* in biological control programs for the tomato moth provided control of over 80%. Oatman and Plantner (1969), studying the cabbage crop in California, USA, found a natural parasitism of *Trichoplusia ni* eggs by microhymenoptera of the genus Trichogramma of around 35%.

Another viable strategy in the management of pests in tomato cultivation is the use of products that are selective to natural enemies (GRAVENA, 1991), which, when associated with the release of species of the Trichogramma genus, allows the control of lepidopteran pests in tomato cultivation to be optimized (CAMPBELL et al., 1991; LEBEDENCO et al., 2007). This association can reduce the number of pesticide applications, providing greater savings and less impact on the environment.

For good results in Integrated Pest Management (IPM) programs, it is important that phytosanitary products that are efficient against pest species are selective to non-target insects (DE GRANDE; GOMEZ, 1990; FALEIRO et al., 1995, REIS et al., 1998). When thinking about integrated production, it is essential that plant protection products are highly efficient against pest species and that they are highly selective to natural enemies, i.e. they do not affect beneficial species such as predators, parasitoids and pathogens (DE GRANDE; GOMEZ 1990; FALEIRO et al. 1995, REIS et al. 1998), minimizing damage to the crop in an economical and environmentally friendly way.

Studies by Siqueira et al. (2000) have shown that constant applications of insecticides to control pests in tomato cultivation cause a number of problems, such as: selection of pest populations resistant to the products used; alteration of the behavior of natural enemies; emergence of pests previously considered secondary, becoming primary; economic and social damage, among others.

Standardized selectivity tests, based on IOBC (International Organization of Biological Control) standards, are necessary for successful field releases. These studies compare the relative toxicity of agrochemicals, including insecticides, fungicides and herbicides, with a view to indicating the most selective products in cases of parasitoid release (HASSAN et al., 1994b; DEGRANDE et al., 2002). Studies on the selectivity of chemical products to Trichogramma species have been carried out using different techniques, such as direct contact with surfaces contaminated with the product, immersion in toxic syrup, residual or persistence analysis and direct spraying (CASTELO BRANCO; FRANÇA 1995, TAKADA et al. 2001, VIEIRA et al. 2001). The most suitable products to be used in pest management are those that combine good control of the pest with the least impact on the activity of natural enemies, and this integration of chemical products with biological control is, in most cases, crucial to the success of agriculture and possible through the use of selective agrochemicals (SANTOS et al., 2006).

1.3 REFERENCES

AGRIANUAL 2012: Yearbook of Brazilian Agriculture: **Tomatoes**. Sao Paulo: FNP,

Consultoria e Agroinformativos, 2012. P. 257.

ALMEIDA, G. D.et al. Effect of azadirachtin on the control of *Anticarsia gemmatalis* and its impact on *Trichogramma pretiosum*. **Phytoparasitica**, v. 38, p. 413-419, 2010.

BLEICHER, E.; PARRA, J.R.P. Species of *Trichogramma* parasitoids of *Alabama argillacea*. III. Determination of the thermal requirements of three populations. **Pesquisa Agropecuària Brasileira,** Brasilia, v.25, n.2, p.215-219, feb. 1990.

BOARETTO, M. A. C.; BRANDÂO, A. L. S., Integrated pest management, Vitória da Conquista, BA, 2000. Universidade Estadual do Sudoeste da Bahia, Departamento de Fitotecnia e Zootecnia Available at: <http://www.uesb.br/entomologia/manejo.html>. Accessed: 26 Nov. 2012.

BOREL, R. M. A.; ABAURRE, M. E. O.; CARMO, C. A. S. do Socio-economic characteristics of tomato cultivation in the state of Espirito Santo. In: TOMATO. Vitória, ES: Incaper, p. 69-84, 2010.

BOTELHO, P.S.M.; PARRA, J.R.P.; CHAGAS NETO, J.F.; OLIVEIRA, C.P.B. Association between the egg parasitoid *Trichogramma galloi* Zucchi (Hymenoptera: Trichogrammatidae) and the larval parasitoid *Cotesia flavipes* (Cam.) (Hymenoptera: Braconidae) in the control of *Diatraea saccharalis* (Fabr.) (Lepdoptera: Crambidae) in sugarcane. **Anais da Sociedade Entomològica do Brasil**, v.28, p.491-496, 1999.

CAMPBELL, R.J.; CAMPBELL, C.W. The "Parvin" mango. Proceedings Florida State. **Horticultural Society**, v.104, p.47-48, 1991.

CARMO, E.L. do; BUENO, A.F.; BUENO, R.C.O.F.; VIEIRA, S.S.;

GOULART,M.M.P.; CARNEIRO, T.R. Selectivity of phytosanitary products used in soybean cultivation for pupae *of Trichogramma pretiosum* Riley, 1879 (Hym.: Trichogrammatidae). **Arquivos do Instituto Biológico**, Sao Paulo, v.77, n.2, p.283-290, 2010.

CARVALHO R.; PERES F. Neoliberalism, the Use of Pesticides and the Crisis of Food Sovereignty in Brazil. In: Breilh J, organizer. **Alternative Report on Health in Latin America.** Quito: CEAS; p. 223-224, 2005.

CARVALHO, C.G.P.; ARIAS, C.A.A.; TOLEDO, J.F.F. et al. Proposed classification of coefficients of variation in relation to soybean yield and plant height. **Pesquisa Agropecuària Brasileira**, v.38, p.187-193, 2003.

CARVALHO, G.A.; PARRA, J.R.P.; BAPTISTA, G.C. Selectivity of some phytosanitary products to two strains of *Trichogramma pretiosum* Riley, 1879 (Hym.: Trichogrammatidae).

Ciência e Agrotecnologia, Lavras, v.25, p.583-591, 2001.

CASTELO BRANCO, M.; FRANÇA, F.H. Impact of insecticides and bioinsecticides on adults of *Trichogramma pretiosum.* **Horticultura Brasileira**, Brasilia, v.13, n.2, p.199-201, 1995.

CORSO, I.; OLIVEIRA, L.J.; FRIEDRICH, R. Effect of insecticides sprayed in the sowing furrow and mixed with soybean seeds on the brown root bug in Cândido Mota, SP. In: WORKSHOP ON BROWN ROOT BUGS, 1., 1999, Londrina. **Minutes and abstracts**. Londrina: Embrapa Soja, p.56-58, 1999.

DEGRANDE, P. E.; P. R. REIS; G. A. CARVALHO; L. C. BELARMINO. Methodology for evaluating the impact of pesticides on natural enemies. p. 75-81 In J. R. P. Parra, P. S. M. Botelho, B. S. Corrêa-Ferreira ; J. M. S. Bento. (Ed.). **Biological Control in Brazil: Parasitoids and predators.** Manole, Sao Paulo. 635 p. 2002.

DEGRANDE, P.E.; GÓMEZ, D.R.S. Selectivity of chemical products in pest control. **Agropecuâria** *Sâo Paulo*, Sao Paulo, v. 7, p. 8-13, I990.

DELPUECH, J.M., C. Dupont, R. Allemand Decrease in fecundity induced by interspecific mating between two Trichogramma parasitoid species. **J. Econ. Entomol**., 103, pp. 308-313, 2010.

FALEIRO, F.G., PICANÇO, M.C.; PAULA, S.U.; BATALHA, V.C. Selectivity of insecticides to *Spodoptera frugiperda* (J.E. Smith) (Lepidoptera: Noctuidae) and to the predator *Dorus luteipes* (Scudder) (Dermaptera: Forficulidae). **Anais da Sociedade Entomológica do Brasil**, v.24, n.2, p.247-252, 1995.

FAO: FAOSTAT. Database Results, Available at: <http://apps.fao.org>. Accessed on: December 19, 2012.

GALLO, D., O. NAKANO, S.S. NETO, R.P.L. CARVALHO, G.C. BATISTA, E.B. FILHO, J.R.P. PARRA, R.A. ZUCCHI, S.B. ALVES, J.D. VENDRAMIM, L.C. MARCHINI, J.R.S. LOPES & C. OMOTO. **Agricultural entomology.** Piracicaba, FEALQ, 920p, 2002.

GIOLO, F.P.; GRÜTZMACHER, A.D.; MANZONI, C.G. *et al.* Toxicity of pesticides used in peach cultivation to immature stages of *Trichogramma pretiosum* Riley (Hym.: Trichogrammatidae). **BioAssay**, v.1, n.4, p.1-7, 2008. Available at: <http://www.bioassay.org.br/articles/1.4/BA1.4.pdf>. Accessed: 26 Nov. 2012.

GOULART R.M.; BORTOLI S.A. DE; THULER R.T.; PRATISSOLI D., VIANNA C.L.T.P., VOLPE X.L.; Evaluation of the selectivity of insecticides to *Trichogramma* spp.

(Hymenoptera: Trichogrammatidae) in different hosts. Arquivos do Instituto Biológico, 75:69-77, 2008.

GRAVENA S. Encontro nacional de produção e abastecimento de tomate, 2. ed. Jaboticabal: FUNEP. p. 105-157. 1991.

GRAVENA, S. Biological control in integrated pest management. **Pesquisa Agropecuària Brasileira**, Brasilia, v.27, n. 4, p.281-299, 1992.

GRAVENA S; BENVENGA SR. 2003. **Practical manual for tomato pest management**. Jaboticabal: Gravena Ltda. 144 p.

HASSAN, S.A., F. BIGLER, H. BOGENSCHUTZ, E. BOLLER, J. BRUN, J.N.M. CALIS, J. COREMANS-PELSENEER, C. DUSO, A. GROVE, U. HEIMBACH, N. HELYER, H. HOKKANEN, G.B. LEWIS, F. MANSOUR, L. MORETH, L. POLGAR, L. SAMS0E-PETERSEN, B. SAUPHANOR, A. STAUBLI, G. STERK, A. VAINIO, M. VAN DE VEIRE, G. VIGGIANI & H. VOGT. Results of the sixth joint pesticide testing program of the IOBC/WPRS - Working Group "Pesticides and Beneficial Organisms". **Entomophaga** 39: 107-119. 1994.

HASSAN, S.A., N. HALSALL, A.P. GRAY, C. KUEHNER, M. MOLL, F.M. BAKKER, J. ROEMBKE, A. YOUSEF, F. NASR & H. ABDELGADER. A laboratory method to evaluate the side effects of plant protection products on *Trichogramma cacoeciae* Marchal (Hym., Trichogrammatidae). p.107-119. In M.P. Candolfi, S. Blümel, R. Forster, F.M. Bakker, C. Grimm, S.A. Hassan, U. Heimbach, M.A. Mead-Briggs, B. Reber, R. Schmuck & H. Vogt. uidelines to evaluate side-effects of plant protection products to non-target arthropods. **IOBC/WPRS.** 2000.

HASSAN, S.A.; H. ABDELGADER. A sequential testing program to assess the effects of pesticides on *Trichogramma cacoeciae* Marchal (Hym., Trichogrammatidae). **IOBC/WPRS Bulletin** 24: 71-81. 2001.

INCAPER, Instituto Capixaba de Pesquisa, Assistência Tècnica e Extensao Rural, **Tomate**, 430p. 2010.

INOUE, M.S.R.; PARRA, J.R.P. Effect of temperature on the parasitism *of Trichogramma pretiosum* Riley, 1879 on eggs of *Sitotroga cerealella* (Olivier, 1819). **Scientia Agricola**, Piracicaba, v.55, n.2, p.222-226, 1998.

ISLA Sementes Ltda. Search for the product Itapua 800 SAIS Long Life Double Hybrid Tomato. Available at: <http://www.isla.com.br/cgi-bin/detalhe.cgi?id=268>. Accessed: 26 Jan. 2013.

KROSS, R.K.; CAVALCANTI MARTA, M.E.R.M.; BRAGA, E.M. Influence of tomato (*Licopersicon esculentum* L.) epidermis on mass transfer during osmotic treatment. Latin American Symposium on Food Science (SLACA), 4. Campinas, **Anais...** Campinas: UNICAMP, 2001.

LEBEDENCO, A., A.M. AUAD & S.N. KRONKA. Methods of lepidopteran control in tomato cultivation (Lycopersicon esculentum Mill.). **Acta Sci. Agron.** 29: 339344. 2007.

LUZ, J.M.Q., SHINZATO A.V.; SILVA M.A.D., Comparison of conventional and organic tomato production systems in protected cultivation. Biosci. J. 23:7 - 15, 2007.

MEDINA, P.; SMAGGHE, G.; BUDIA, F. *et al.* Toxicity and absorption of azadirachtin, diflubenzuron, pyriproxyfen, and tebufenozide after topical application in predatory larvae of *Chysoperla carnea* (Neurop.: Chrysopidae). **Environmental Entomology**, v.32, p.196-203, 2003.

MELO, P.C.T. Desenvolvimento sustentâvel da cadeia produtiva do tomate para consumo in natura no Brasil e os desafios do melhoramento genético. Horticultura Brasileira, v21, n.2, July, 2003.

MOLINA, R. M. S.; V. FRONZA & J. R. P. PARRA. Selection of *Trichogramma* spp. for the control of *Ecdytolopha aurantiana* based on biology and thermal requirements. **Revista Brasileira de Entomologia 49**: 151-158. 2005.

NAVA, D.E.; NACHTIGAL, G.F. Biological control in southern Brazil. In: G.BIO **- Revista de Controle Biológico**, Piracicaba: Facile, p.15-18. 2010.

NORNBERG, S.D.; GRÜTZMACHER, A.D.; KOVALESKI, A.; CAMARGO, E.S.;PASINI, R.A. Toxicity of pesticides used in integrated apple production to Trichogramma pretiosum Riley, 1879 (Hym.: Trichogrammatidae) under laboratory conditions. **Revista Brasileira de Agrociência**, Pelotas, v.15, n.1 - 4, p.49 - 56, 2009.

OATMAN, E.R.; G.R. PLATNER. An ecological study of insect populations on cabbage in southern California. **Hilgardia** 40:1-40. 1969.

PARRA, J. R. P. Applied biological control and integrated pest management. In: SIMPÓSIO DE AGRICULTURA ECOLÓGICA, 1, Campinas, 1993. Campinas: **Cargill Foundation**, p.116-139. 1993.

PRATISSOLI Dirceu, FORNAZIER M.J., HOLTZ A.M., GONÇALVES JR., CHIORAMITAL A.B. & ZACO H.B.; Occurrence of *Trichogramma pretiosum* in commercial tomato areas, in Espirito Santo, in regions of different altitudes. **Horticultura Brasileira**, 21:73-76. 2003.

PRATISSOLI, Dirceu, F.F. PEREIRA, R. BARROS, J.R.P. PARRA & C.L.T. PEREIRA. Parasitism *of Trichogramma pretiosum* on crucifer moth eggs under different temperatures. **Hortic. Bras.** 22: 754-757. 2004.

PRATISSOLI, D.; PARRA, J.R.P. Development and thermal requirements of *Trichogramma pretiosum* (Riley), reared on two tomato moths. **Pesquisa Agropecuària Brasileira,** v.35, p.1281-1288, 2000.

PROTAS, J. F. da S.; KREUZ, C. L.; FREIRE, J. de M. Less inputs, more quality. **Cultivar Hortaliças e Frutas**, Pelotas, v. 3, n. 17, p. 20-24, jan. 2003.

REIS, P.R., L.G. CHIAVEGATO, G.J. MORAES, E.B. ALVES & E.O. SOUSA. Selectivity of agrochemicals to the predatory mite *Iphiseiodes zuluagai* Denmark & Muma (Acari: Phytoseiidae). An. **Soc. Entomol. Brasil** 27:265-274. 1998.

RUBERSON, J.R.; TILLMAN, P.G. Effect of selected insecticides on natural enemies in cotton: laboratory studies, In: **Proceedings. Beltwide Cotton Conference**, January 3-7, 1999, New Orleans, LA. National Cotton Council, Orlando, Florida. p.12101213. 1999.

SA, L.A.N.; PARRA, J.R.P. Biology and parasitism of Trichogramma pretiosum (Riley) (Hym.; Pyralidae) on Ephestia kuehniella (Zeller) and Heliothis zea (Boddie) (Lep.: Noctuidae) eggs. **Journal of Applied Entomology**, v.118, p.38-43, 1994.

SANTOS, A. C.; BUENO, A. F.; BUENO, R. C. O. Selectivity of pesticides to natural enemies. In: PINTO. A. S.; NAVA, D. E.; ROSSI, M. M. MALERBO-SOUZA, D. T. (Eds.). **Biological pest control: in practice**, Pinto: Piracicaba. Chap.18, p 221-227, 2006

SECRETARIA DE ESTADO DA AGRICULTURA, ABASTECIMENTO, AQUICULTURA E PESCA DO ESTADO DO ESPÌRITO SANTO, A **tomaticuitura capixaba em destaque.** <http://www.seag.es.gov.br/?p=13978> Accessed on: November 23, 2012.

sIQuEIRA, H.A.A.; GuEDEs, R.N.c.; PIcANÇo, M.c. Insecticide resistance in populations of Tuta absoluta (Lepidóptera: Gelechiidae). **Agricultural and Forest Entomology,** v.2, n.2, 147-153, 2000.

souZA, J.c.; REIs, P.R. Main pests of table tomatoes: bioecology, damage and control. **Informe Agropecuârio**, v.24, p.79-92, 2003.

TAKADA, Y.; KAWAMuRA, s.; TANAKA, T. Effects of various insecticides on the development of the egg parasitoid *Trichogramma dendrolimi* (Hymenoptera: Trichogrammatidae). **Journal of Economic Entomology**, Lanham, v.94, n.6, p.1340 - 1343, 2001.

TAMISO, L.G.; Master's thesis: Performance of tomato cultivars (*Lycopersicon esculentum* Mill.) under organic systems in protected cultivation. University of São Paulo, ESALQ, Piracicaba, p. 12 - 14, 2005.

VALE, F.X.R.; JESUS JÛNIOR, W.C.; LIBERATO, J.R.; SOUZA, C.A. Quantification of diseases and host growth. In: VALE, F.X.R.; JESUS JÛNIOR, W.c.; ZAMBOLIM, L (Ed.). **Epidemiology applied to plant disease management.** Belo Horizonte: Perffil, 2004. v.1, chap.3, p.91-126.

VIEIRA, A.; OLIVEIRA, L.; GARclA, P. Effects of conventional pesticides on the preimaginal developmental stages and on adults *Trichogramma cordubensis* (Hymenoptera: Trichogrammatidae). **Biocontrol Science and Technology**, v.11, n.4, p.527-534, 2001.

ZUccHI, R.A, MONTEIRO, R.c. The genus Trichogramma in South America. In: PARRA, J.R.P.; ZUccHI, R.A. (Ed.). **Trichogramma and applied biological control.** Piracicaba: FEALQ, 1997. p.41-46.

CHAPTER II

ACTION OF INSECTICIDES USED ON tomato CROPS on *trichogramma pretiosum* riley, 1879 (hymenoptera: trichogrammatidae).

REsuMo

Biological control and the use of insecticides need to be made compatible in order to preserve the community of beneficial insects in the production system. Therefore, the aim of this work was to evaluate the changes in the biological characteristics of *Trichogramma pretiossum* (Riley, 1879) (Hymenoptera: Trichogrammatidae), when parasitized by *Anagasta kuehniella* (Zeller, 1879) (Lepidoptera: Pyralidae) eggs treated with insecticides recommended for tomato cultivation for selectivity tests. The experiment was conducted at the Núcleo de Desenvolvimento Cientifico e Tecnolòico em Manejo Fitossanitârio (NUDEMAFI), in air-conditioned chambers set to a temperature of 25±1°C, relative humidity of 70±10%, and 14-hour photophase. One newly emerged female of *T. pretiosum* was used in each Eppendorf tube containing a carton with 20 inviable eggs of the host *A. kuehniella*, which had been previously emerged in insecticide syrup. There were 15 replicates for each treatment. The eggs were parasitized for 24 hours and after the offspring emerged, the biological characteristics of the parasitism and emergence of the individuals were checked. Insecticides with the active ingredient PIRIPROXIFEM, TEFLUBENZURON, ACETAMIPRID 1, LAMBDA-CIALOTRINA and CHLORANTRANILIPROLE + LAMBDA-CIALOTRINA were considered the most selective for the parasitoid *T. pretiosum*.

Keywords: Biological control. Phytosanitary products. Selectivity.

2.1 INTRODUCTION

The social importance of tomatoes lies in the fact that the activity is highly labor-intensive, given that for every hectare occupied, there are on average five rural workers (SEAG-ES, 2012). The technological level of the crop is considered to be very high, since it requires high investments in inputs and due to the high risk, and because of the high infestation of pests and incidence of diseases, both in crops destined *for fresh* consumption and for industry (SOUZA; REIS, 2003; LUZ et al., 2007).

Pest management can be defined as the joint use of techniques aimed at controlling phytosanitary disorders, analyzing the cost/benefit ratio in line with the interests and/or impact on producers, society and the environment (GALLO et al. 2002, PICANÇO et al. 2004, VALE et al. 2004). Therefore, in general, both agriculture and tomato growing should seek to use methods that aim to be rational and efficient in order to stimulate biological processes and recover the potential of environmental biodiversity (PROTAS, 2003).

It is of great importance to study and improve pest management methodologies that aim to integrate techniques, making the use of chemicals compatible, since this is the method most used in conventional tomato growing, although efficient in most cases, pesticides can negatively affect the community of natural enemies (CARMO et al. 2010).

Biological control has been shown to be one of the options for association with other methods, especially with egg parasitoids such as those of the genus Trichogramma (PRATISSOLI et al., 2004), which has 210 known species worldwide, distributed in all agricultural environments on the planet (PINTO, 2006). These parasitoids are capable of parasitizing more than 200 species of insects that cause economic damage, mainly Lepidoptera (PRATISSOLI et al., 2004), in which they cause the death of the developing embryos (JARJEES; MERRITT, 2004).

In order to make biological and chemical control compatible, the aim of this study was to evaluate the biological characteristics of *Trichogramma pretiosum* when parasitized on eggs treated with insecticides, as well as to classify insecticides in terms of their impact on natural enemies, according to the recommendations of the IOBC/WPRS.

2.2 MATERIAL AND METHODS

The selectivity and susceptibility experiments were carried out in the Entomology sector of the Núcleo de Desenvolvimento Cientifico e Tecnolòico em Manejo Fitossanitâ (NUDEMAFI) of the Centro de Ciências Agrârias da Universidade Federal do Espirito Santo (CCA-UFES), in Alegre - ES. All the experiments were carried out in an air-conditioned chamber with the following characteristics: temperature 25 ± 1 °C, relative humidity (RH) 70 ± 10% and 14-hour photophase.

2.2.1 Breeding and maintenance of *Trichogramma pretiosum*

To maintain and multiply the parasitoid *T. pretiosum*, the alternative host *A. kuehniella* was used, according to the methodology developed by Pratissoli et al. (2010a). To maintain the host, a diet based on whole wheat flour (70% m/m), corn flour (27% m/m) and brewer's yeast (3% m/m) was used.

To raise the parasitoid, eggs from the host were adhered to blue cardboard cards (8 x 2 cm) using gum arabic diluted to 50% (w/v). These eggs were then submitted to impregnation under a germicidal lamp for 50 minutes, as described by Pratissoli et al. (2010a), and then exposed to parasitism for a period of 24 hours and kept in an air-conditioned chamber at

25±1°C, RH of 70±10% and photophase of 14 hours.

2.2.2 Selectivity bioassay

The choice of products took into account the volume sold in specialized retailers in the tomato producing region of Espirito Santo, the commercial availability of the products and registrations with the Espirito Santo Agricultural and Forestry Defense Institute (IDAF). 30 commercial insecticides were tested (Table 1).

Table 1. Products evaluated in the action of insecticides used on tomato crops on *Trichogramma pretiosum*.

(continued)

Technical name	[ ] i.a.*	Recommended application dose	Dose applied in 200 ml
ABAMECTIN	18 g/L	up to 100 mL/100 L of water	0.2 mL
ACEPHATE	750 g/kg	up to 100 g/100 L of water	0,2 g
ACETAMIPRID 2	200 g/kg	250 to 400 g/ha	2,67 g
ACETAMIPRID 1	200 g/kg	250 to 400 g/ha	0,08 g
ALPHA-CYPERMETHRIN	100 g/L	up to 10 mL/100 L of water	0.02 mL
BETA-CYFLUTHRIN	50 g/L	up to 25 mL/100 L of water	0.05 mL
BETA-CYFLUTHRIN + IMIDACLOPRID	12,5 + 100 g/L	750 to 1000 mL/ha	0.2 mL
BUPROFEZIN	250 g/kg	100 to 200 g/100 L of water	0,4 g
CARBOFURAN	350 g/L	up to 5 L/100 L of water	10 mL
CARBOSULPHANE	400 g/L	up to 50 mL/100 L	0.1 mL
CHLORANTRANILIPROLE	200 g/L	up to 200 mL/ha	0.02 mL
CHLORANTRANILIPROLE + LAMBDA-CYHALOTHRIN	100 + 50 g/L	20 to 30 mL/100 L of water	0.06 mL
CHLORFENAPYR	240 g/L	25 to 50 mL/100 L of water	0.1 mL
CYPERMETHRIN + PROFENOFOS	40 + 400 g/L	up to 125 mL/100 L of water	0.25 mL
CYROMAZINE CHLORIDRATE	750 g/kg	up to 15 g/100 L of water	0,03 g
CARTAPE	500 g/kg	up to 250 g/100 L of water	0,5 g
CHLORPYRIPHOS	480 g/L	up to 1.5 L/ha	0.3 mL
DELTAMETRINE	25 g/L	up to 50 mL/100 L of water	0.1 mL
SPIROMESIPHENE	240 g/L	500 to 600 mL/ha	0.12 mL
ETOFENPROX	300 g/L	up to 60 mL/100 L of water	0.12 mL
FENPROPATHRINE	300 g/L	up to 150 mL/ha	0.04 mL
INDOXACARB	300 g/kg	up to 16 g/100 L of water	0,032 g
LAMBDA-CYHALOTHRIN	50 g/L	40 to 50 mL/100 L of water	0.1 mL
LAMBDA-CYHALOTHRIN + THIAMETHOXAM	106 + 141 g/L	50 to 100 mL/100 L of water	0.2 mL
METHOMYL + METHOMYL	215 + 215 g/L	up to 100 mL/100 L of water	0.2 mL
MILBEMECTIN	50 g/L	up to 40 mL/100 L of water	0.06 mL
PIRIPROXIFEM	100 g/L	75 to 100 mL/100 L of water	0.2 mL
TEFLUBENZURON	150 g/L	up to 25 mL/100 L of water	0,05
THIAMETHOXAM	250 g/kg	16 to 20 g/100 L of water	0,04 g
TRIFLUMURON	480 g/L	up to 30 mL/100 L of water	0.06 mL

* Concentration of the active ingredient in the commercial product.

Newly emerged females *of T. pretiosum* (age < 24 hours) were individualized in Eppendorf tubes (2.5 ml) containing a drop of honey inside for feeding. Sky blue cardboard cards (0.5 x 3.0 cm) containing 20 eggs of the host *A. kuehniella* glued together with gum arabic (25%

m/v) were rendered unviable by exposure to a germicidal lamp for approximately 50 minutes.

The cartons containing the eggs were immersed in the product syrups for a period of five seconds. In the control treatment, only distilled water was used. The cards were then placed on paper towels in plastic trays in an exhaustion chamber for one hour at room temperature to remove excess moisture from the surface of the eggs. The cartons containing the treated eggs were offered to parasitized *T. pretiosum* for 24 hours. After this period, the females were removed and the cards were kept in the same tubes for observation until the parasitoids emerged.

The parameters observed were the percentage of parasitism. Each treatment consisted of 15 replicates represented by a card containing 20 eggs of the host under study.

2.2.3 Statistical analysis

The design used was entirely randomized with 31 treatments (30 insecticides + control) and 15 replicates each. The parasitism data obtained in the bioassays was submitted to the parasitism reduction formula and then to analysis of variance (ANOVA) and the means were compared using the Scott-Knott test at 5% probability. The statistical analysis was carried out using the statistical program SISVAR version 5.1 (FERREIRA, 2003).

The reduction in parasitism of the parasitoid for each treatment was determined by comparing it with the control and calculated using the formula: RP = (1 - Rt/Rc) x 100, where RP is the percentage reduction in parasitism; Rt is the average parasitism value for each product and Rc is the average parasitism observed for the control treatment (HASSAN et al., 2000). Based on these reduction percentages, the products tested were classified into classes, according to the IOBC/WPRS recommendations (1988 and 1992), where: 1 = harmless (< 30%), 2 = slightly harmful (30 - 79%), 3 = moderately harmful (80 - 99%) and 4 = harmful (> 99%).

2.3 - Results and Discussion

The parasitism of *T. pretiosum* was affected by those evaluated (F=23.46; glr=237; p<0.001) (Table 2).

Table 2. Percentage of parasitism, reduction in parasitism (RP) *of T. pretiosum* in relation to the control and IOBC/WPRS classes of insecticides recommended for tomato cultivation.

Treatment	Parasitism (%) *	PR (%)[1]	IOBC Class[2]
ACEPHATE	82,14 a	(29,93) **	1
CHLORANTRANILIPROLE + LAMBDA-CYLOTRIN	81,00 a	(29,11) **	1
ALPHA-CYPERMETHRIN	80,00 a	(27,52) **	1
CYROMAZINE	75,66 a	(20,61) **	1

TRIFLUMURON	75,50 a	(20,35) **	1
CHLORFENAPYR	74,66 a	(19,02) **	1
CHLORANTRANILIPROLE	74,64 a	(18,98) **	1
LAMBDA-CYHALOTHRIN	70,90 a	(13,03) **	1
ACETAMIPRID 1	67,22 a	(7,15) **	1
TEFLUBENZURON	63,07 b	(0,54) **	1
PIRIPROXIFEM	61,87 b	1,36	1
BUPROFEZIN	57,30 b	8,64	1
INDOXACARB	52,50 b	16,31	1
METOMIL	50,55 b	19,41	1
ABAMECTIN	45,00 b	28,26	1
ACETAMIPRID 2	33,33 c	46,86	2
BETA-CYFLUTHRIN	25,66 c	59,08	2
SPIROMESIPHENE	22,14 c	64,7	2
BETA-CYFLUTHRIN + IMIDACLOPRID	13,18 d	78,98	2
ETOFENPROX	11,66 d	81,4	3
FENPROPATHRINE	7,50 d	88,04	3
CYPERMETHRIN + PROFENOFOS	7,00 d	88,84	3
CARBOFURAN	-- e	100,00	4
CARBOSULPHANE	-- e	100,00	4
CARTAPE HYDROCHLORIDE	-- e	100,00	4
CHLORPYRIPHOS	-- e	100,00	4
DELTAMETRINE	-- e	100,00	4
LAMBDA-CYHALOTHRIN + THIAMETHOXAM	-- e	100,00	4
MILBEMECTIN	-- e	100,00	4
WITNESS	50,72 b	---	---
CV (%)	28,47		
F	23,46		
GLR	237		
P	<0,001		

* Averages followed by the same letter do not differ statistically by the Scott-Knott test at 5% probability;

** Parasitism Reduction (PR) values in brackets indicate an increase in parasitism compared to the control;

[1] PR = Reduction in parasitism compared to the control.

[2] IOBC/WPRS classes for selectivity testing on Trichogramma: 1 = innocuous (< 30%), 2 = slightly harmful (30-79%), 3 = moderately harmful (80-99%), 4 = harmful (> 99%).

The action of insecticides can be influenced by the molecule, concentration or dose applied or by the biological organisms (parasitoid and host). It was observed that the parasitoid did not accept the host when they were immersed in some insecticide syrups. This may be due to contact with the antennae or even with the ovipositor itself and walking during the evaluation, which caused the potential host to be rejected, preventing oviposition or subsequently inhibiting the parasitoid's development (MOREIRA et al. 2009). Based on the parasitism data analyzed, statistical differences were observed between the active ingredients of the insecticides, which made it possible to subdivide them into five distinct groups and four classes of impact on natural enemies, according to the IOBC/WPRS recommendations.

The results shown in Table 2 show that the insecticides based on ACEPHATE, CHLORANTRANILIPROLE + LAMBDA-CIALOTRIN, ALPHA- CYPERMETHRIN,

CIROMAZINE, TRIFLUMURON, CHLORFENAPYR, CHLORANTRANILIPROLE, THIAMETHOXAM, LAMBDA-CIALOTRIN, ACETAMIPRID 1 and TEFLUBENZURON promoted an increase in average parasitism when compared to the control, characterizing the phenomenon known as hormesis, which is defined as the stimulation of an organism's performance by small exposures to harmful or toxic agents at high levels (MAGALHAES et al., 2002; ZANÙNCIO et al., 2003; CALABRESE et al., 2008). This stimulatory phenomenon of overcompensation was suggested by Thownsend and subsequently by other authors (FORBES 2000; CALABRESE 2008).

The best results among the products tested were represented by the insecticides ACEPHATE, CHLORANTRANILIPROLE + LAMBDA-CYALOTHRIN, ALPHA-CYPERMETHRIN, CIROMAZINE, TRIFLUMURON, CHLORFENAPYR, CHLORANTRANILIPROLE, THIAMETHOXAM, LAMBDA-CIALOTRIN and ACETAMIPRID 1, which did not differ statistically from each other by the grouping of means test, showed a variation in parasitism between 82.14 and 67.22%, obtaining an increase in the reduction of parasitism when compared to the control and were classified as innocuous by the IOBC/WPRS classification (Table 2).

The active ingredients TEFLUBENZURON, PIRIPROXIFEM, BUPROFEZIN, INDOXACARB, METOMIL and ABAMECTIN have been grouped together as insecticides

Intermediates, which showed parasitism values between 63.07 and 45.00% and promoted a reduction in parasitism between 0.54 and 28.26%, were classified as innocuous (class 1) by the IOBC/WPRS.

For the insecticides with the active ingredient ACETAMIPRID 2, BETA-CYFLUTHRIN and ESPIROMESIFENO, the parasitism observed did not differ statistically, where parasitism ranging from 33.33 to 22.14% was observed, with parasitism reduction values ranging from 46.86 to 64.7% and classified as slightly harmful to the parasitoid *T. pretiosum*, classified in the IOBC/WPRS table as class 2.

For the insecticides with the active ingredient BETA CYFLUTHRIN + IMIDACLOPRIDO, ETOFENPROX, FENPROPATHRINA and CIPERMETRINA + PROFENOFÓS, no differences were observed in parasitism, which ranged from 13.18 to 7.00%,00% and promoted a reduction in parasitism ranging from 78.98 to 88.84%, these being classified according to the IOBC/WPRS as moderately harmful except for the insecticide with the active ingredient BETA-CYFLUTHRIN + IMIDACLOPRIDO which was classified as slightly harmful.

Insecticides with the active ingredients CARBOFURAN, CARBOSULPHAN, CARTAPE CHLORIDRATE, CHLORPYRIPHOS, DELTAMETHRIN, LAMBDA-CYALOTHRIN + THIAMETHOXAM and MILBEMECTIN did not parasitize the parasitoid *T.* pretiosum. *pretiosum*, a fact which implies that parasitism was not observed on the eggs of the host. These reasons may be related to the influence of the molecule, concentration or dose applied to the fungicides or even the biological organisms involved in the bioassays (parasitoid and host). For the purposes of the IOBC/WPRS classification, they were classified as harmful (class 4) to the egg parasitoid *T. pretiosum.*

The TRIFLUMURON-based insecticide showed favorable results for parasitism, results that corroborate those found by CARVALHO et al. (2003) in selectivity studies with *T. pretiosum* on tomato. The results found for ABAMECTIN were similar to those found by CARVALHO et al. (2003), in studies of insecticide selectivity in tomato crops with *T. pretiosum* and WANG et al. (2011) who worked on the toxic effects of insecticides on *Trichogramma ostriniae* (Hymenoptare: Trichogrammatidae), where ABAMECTIN caused a significant reduction in parasitism and emergence in all immature stages of *T. pretiosum.*

The negative action of CYPERMETHRIN + PROPHENOPHOS found in the selectivity bioassays was discussed by Maich et al. (2011), where studies of the selectivity of organophosphate insecticides in soybean crops on the parasitoid *T. pretiosum* showed that this active ingredient is not selective to the parasitoid, as there were no parasitized eggs, so they concluded that the active ingredient in question is harmful to this species of Trichogramma.

The organophosphates (CHLORPYRIPHOS, ACEPHATE and CYPERMETHRIN + PROPHENOPHOS) used in the selectivity bioassays found in Table 2 were classified as moderate and highly harmful to *T. pretiosum,* results which are similar to those found by WANG et al (2012a) and ZHAO et al. (2012) who conducted laboratory toxicity studies of insecticides on *A. kuehniella* eggs parasitized with *T. pretiosum.*

The insecticides with the principle ACETAMIPRID 1 and CHLORFENAPYR obtained similar values in terms of the average parasitism observed, 74.66 and 67.22% respectively, where they promoted an increase in the reduction of parasitism with values of (19.02%) and (7.15%), being classified as innocuous by the IOBC/WPRS. These data corroborate the data obtained by Moura et al. (2005) who reported on the toxicity of insecticides used on tomato crops based on ACETAMIPRID 1, CHLORFENAPYR and TIAMETOXAM on *T. pretiosum* where they concluded that ACETAMIPRID and TIAMETOXAM are harmless to the parasitoid's adults.

2.4 CONCLUSION

The parasitism of the parasitoid *T. pretiosum* was affected by the insecticides used on the tomato crop, where the insecticides with the active ingredient ACEPHATE, CHLORANTRANILIPROLE + LAMBDA-CIALOTRIN, ALPHA- CYPERMETHRIN, CIROMAZINE, TRIFLUMURON, CHLORFENAPYR, CHLORANTRANILIPROLE, THIAMETHOXAM, LAMBDA-CIALOTRIN, ACETAMIPRID 1 and TEFLUBENZURON promoted an increase in average parasitism when compared to the control group. This group is considered the most selective because it makes biological and chemical control compatible.

2.5 - REFERENCES

CALABRESE E.J., Converging concepts: Adaptive response, preconditioning, and the Yerkes-Dodson Law are manifestations of hormesis. **Ageing Research Reviews**, 7, 8-20, 2008.

CARVALHO, G. A.; PARRA, J. R. P.; BAPTISTA, G. C. Bioactivity of phytosanitary products used in tomato cultivation (*Lycopersicon esculentum* MILL.) to *Trichogramma pretiosum* Riley, 1879 (Hymenoptera: Trichogrammatidae) in F1 and F2 generations. **Ciência e Agrotecnologia**, v. 27, n. 2, p. 261-270, 2003.

DELPUECH JM, BARDON C.; BOULÉTREAU M., Increase of the behavioral response to kairomones by the parasitoid wasp *Leptopilina heterotoma* surviving insecticides. Archives Environmental Contamination Toxicology, 49:186-191, 2005.

FERREIRA, D.F. **SISVAR**: version 5.1. Lavras: UFLA, 2003.

FORBES, V.E., Is hormesis an evolutionary expectation? **Functional Ecology**, 14, 12-24, 2000.

GIOLO, F. P.; GRÜTZMACHER, A. D.; MANZONI, C. G.; LIMA, C. A. B. NORNBERG, S. D. Toxicity of phytosanitary products used in peach cultivation on adults of *Trichogramma pretiosum*. **Bragantia**, v. 66, p.423-431, 2007.

COULART, R.M.; BORTOLI, S.A.; THULER, de, R.T., D. PRATISSOLI, VIANA, C.L.T.P.; H.X.L. VOLPE. Evaluation of the selectivity of insecticides to *trichogramma* spp. (hymenoptera: trichogrammatidae) in different hosts. **Archive: Biological Institute of Sao Paulo**. Sao Paulo, v.75, n.1, p.69-77, jan./mar., 2008.

HASSAN, S.A., N. HALSALL, A.P. GRAY, C. KUEHNER, M. MOLL, F.M. BAKKER, J. ROEMBKE, A. YOUSEF, F. NASR & H. ABDELGADER. A laboratory method to evaluate the side effects of plant protection products on *Trichogramma cacoeciae* Marchal (Hym. Trichogrammatidae). p.107-119. In M.P. Candolfi, S. Blümel, R. Forster, F.M. Bakker, C. Grimm, S.A. Hassan, U. Heimbach, M.A. Mead-Briggs, B. Reber, R. Schmuck & H. Vogt. uidelines to evaluate side-effects of plant protection products to non-target arthropods. **IOBC/WPRS.** 2000.

HASSAN, S.A.; HAFES, H.; DEGRANDE, P.E.; HERAI, H. The side effects of pesticides on the egg parasitoid *Trichogramma caecoeciae* Marchal (Hym., Trichogrammatidae), acute dose-response and persistence tests. **Journal of Applied Entomology**, v. 122, n. 9-10, p. 569-573, 1998.

JARJEES, E.A.; MERRITT, D.J. The effect of parasitization by *Trichogramma australicum* on *Helicoverpa armigera* host eggs and embryos. **Journal of Invertebrate Pathology**, v. 85, p. 1-8, 2004.

LUZ, J. M.Q.; SHINZATO, A. V.; SILVA, M. A. D. Comparison of conventional and organic tomato production systems in protected cultivation. **Bioscience Journal**, v. 23, p. 7-15, 2007.

MAGALHAES, L.C.; GUEDES, R.N.C.; OLIVEIRA, E.E.; TUELHER, E.S. Development and reproduction of the predator Podisus distinctus (Stal) (Heteroptera: Pentatomidae) against sublethal doses of permethrin. Crop Protection, 31, 445-448, 2002.

MAICH, S. L. S. da P.; MAGANO, D. A.; ZIMMER, M.; PAULUS, L. F.; GRÜTZMACHER, A. D.; selectivity of organophosphate insecticides registered for caterpillars associated with soybean crops on the egg parasitoid *trichogramma pretiosum* Riley, 1879 (hymenoptera: trichogrammatidae); XX Congresso de iniciaçâo cientifica - 3° Mostra Cientifica. Federal University of Pelotas, 2011.

MOURA, A. P.; CARVALHO, G. A.; OLIVEIRA, R. L. de. Toxicity of insecticides used in tomato cultivation to *Trichogramma pretiosum.* **Pesq. agropec. bras.**, Brasilia, v.40, n.3, p.203-210, mar. 2005

MOURA, A. P.; CARVALHO, RIGITANO, G. A.; R. L. de O. Toxicity of insecticides used on tomato crops to *Trichogramma pretiosum.* **Pesquisa Agropecuària Brasileira**, Brasilia, v.40, n.3, p.203-210, mar. 2005.

PINTO, J.D. A review of the New World genera of Trichogrammatidae (Hymenoptera). **Journal of Hymenoptera Research**, Washington, v.15, p. 38-163, 2006.

PRATISSOLI, D., HOLTZ, A. M.; GONÇALVES, J. R.; VIANNA, U. R.; BELLINI, L. L., Effect of absence of host and food on biological aspects of *Trichogramma pretiosum* Riley (Hymenoptera: Trichogrammatidae). **Acta Scientiarum Biological Science**, Maringà, v. 26, p. 281-286, 2004.

PRATISSOLI, D.; DALVI, L. P.; POLANCZYK, R. A.; ANDRADE, G. S.; HOLTZ, A. M.; NICOLINE, H. O. Biological characteristics of *Trichogramma exiguum* on eggs of *Anagasta kuehniella* and *Sitotroga cerealella*. **Idésia**, v. 28, n. 1, p. 39-42, 2010a.

PRATISSOLI, D.; FORNAZIER, M.J.; HOLTZ, A.M.; GONÇALVES, J.R.; CHIORAMITAL, A.B.; ZAGO, H. Occurrence of *Trichogramma pretiosum* in commercial tomato areas in Espirito Santo, in regions of different altitudes. **Horticultura Brasileira,** Brasilia, v. 21, n. 1, p. 73-76, 2002.

PRATISSOLI, D.; MILANEZ, A. M.; BARBOSA, W. F.; CELESTINO, F. N.; ANDRADE, G. S.; POLANCZYK, R. A. SIDE EFFECTS OF FUNGICIDES USED IN CUCURBITACEOUS CROP ON *Trichogramma atopovirilia* OATMAN & PLATNER (HYMENOPTERA: TRICHOGRAMATIDAE). **Chilean Journal of Agricultural Research**, v. 70, n. 2, p. 323-327, 2010b.

PRATISSOLI, D.; MILANEZ, A. M.; CELESTINO, F. N.; BARBOSA, W. F.; VIANNA, U. R.; POLANCZYK, R. A.; ZINGER, F. D.; CARVALHO J. R. de. Selectivity of insecticides recommended for cucurbits for *Trichogramma atopovirilia* Oatman & Platner (Hymenoptera: Trichogrammatidae) under laboratory conditions. **Rev. Ceres**, Viçosa, v. 58, n.5, p. 661-664, Sep/Oct, 2011.

PRATISSOLI, D.; VIANNA, U. R.; FURTADO, G. O.; ZANUNCIO, J. C.; POLANCZYK, R. A.; BARBOSA, W. F.; CARVALHO, J. R. Selectivity of insecticides to *Trichogramma pretiosum* Riley (Hymenoptera: Trichogrammatidae) in different hosts. **Boletin de Sanidad Vegetal Plagas**, v. 35, n. 3, p. 347-353, 2009.

PRATISSOLI, D; PARRA, J.R.P. Selection of *Trichogramma pretiosum* Riley (Hymenoptera: Trichogrammatidae) Strains for the Control of the Moths *Tuta absoluta* (Meyrick) and *Phthorimaea operculella* (Zeller) (Lepidoptera: Gelechiidae). **Neotropical Entomology**, Londrina, v. 30, n. 2, p. 277-282, 2001.

QUERINO, R.B.; ZUCCHI, R.A.; PINTO, J.D. Systematics of the Trichogrammatidae (Hymenoptera: Chalcidoidea) with a Focus on the Genera Attacking Lepidoptera. In: CÔNSOLI, F.L.; PARRA, J.R.P.; ZUCCHI, R.A. (Eds.) **Egg Parasitoids in Agroecosystems with Emphasis on *Trichogramma***. Piracicaba: Springer, p. 101218, 2010.

RUSSELL, J. E.; STOUTHAMER, R. Sex Ratio Modulators of Egg Parasitoids. In: CÔNSOLI, F. L.; PARRA, J. R. P.; ZUCCHI, R. (Eds.) Egg Parasitoids in Agroecosystems with Emphasis on *Trichogramma*. **Springer**, Piracicaba, p. 167-190, 2010.

SCHOLZ B.C.G., ZALUCKI M.P., The effects of two new insecticides on the survival of adult Trichogramma pretiosum Riley in sweet corn. In: Austin A, Dowton M, editors. **Hymenoptera: Evolution, Biodiversity and Biological Control.** Canberra, Australia: CSIRO Pubs. p. 381-387, 2000.

SOUZA, J.C.; REIS, P.R. Main pests of table tomatoes: bioecology, damage and control. **Informe Agropecuàrio**, v.24, p.79-92, 2003.

TROUGAKOS, I., L. MARGARITIS. Novel morphological and physiological aspects of insect eggs. In M. Hilker and T. Meiners (eds.). Chemoecology of insect eggs and egg deposition. **Blackwell Publishing Ltd**, Berlin, Germany, p3-36, 2002.

VIANNA, U.R. et al. Insecticide toxicity to *Trichogramma pretiosum* (Hymenoptera: Trichogrammatidae) females and effect on descendant generation. **Ecotoxicology**, v.18, n.2, p. 180-186, 2009.

VINSON, S. B., Host selection behavior of egg parasitoids, with emphasis on the Trichogrammatidae family, 67-120. In: PARRA, J. R. P., ZUCCHI, R. A.

(eds.). *Trichogramma* **and applied biological control**. Piracicaba: FEALQ, 324pp. 1997.

WAJNBERG, E. Genetics of the Behavioral Ecology of Egg Parasitoids. In: CÔNSOLI, F. L.; PARRA, J. R. P.; ZUCCHI, R. (Eds.) Egg Parasitoids in Agroecosystems with Emphasis on *Trichogramma*. **Springer**, Piracicaba, p. 149-166, 2010.

WANG, Y. H., L. P. CHEN, R. X. RUI, X. P. ZHAO, C. X. WU, T. CANG; Q. WANG. Insecticide toxic effects on *Trichogramma ostriniae* (Hymenoptera: Trichogrammatidae). **Pest Manage. Sci.**, 68: 1564-1571, 2012a.

ZANUNCIO, T.V., SERRÂO, J. E.; ZANUNCIO, J. C.; GUEDES, R. N. C. Permethrin-induced hormesis on the predator *Supputius cincticeps* (Stal, 1860) (Heteroptera: Pentatomidae). **Crop Protection**. v. 22, p. 941-947, 2003.

ZANUNCIO, T.V., SERRÂO, J.E., ZANUNCIO, J.C.; GUEDES, R.N.C., Permethrin- induced hormesis on the predator *Supputius cincticeps* (Stal, 1860) (Heteroptera: Pentatomidae). **Crop Protection**, 22, 941-947, 2003.

ZHAO, X. P., C. X. WU, Y. H. WANG, T. CANG, L. P. CHEN, R. X. RUI; Q. WANG. Assessment of toxicity risk of insecticides used in rice ecosystem on *Trichogramma japonicum,* an egg parasitoid of rice Lepidopterans. **J. Econ. Entomol.** 105: 92-101, 2012.

CHAPTER III

THE ACTION OF FUNGICIDES USED ON tomato CROPS on *trichogramma pretiosum* riley, 1879 (hymenoptera: trichogrammatidae).

REsuMo

The aim of this work was to evaluate the possible changes in the biological characteristics of *Trichogramma pretiossum* (Riley, 1879) (Hymenoptera: Trichogrammatidae) when parasitized by *Anagasta kuehniella* (Zeller, 1879) (Lepidoptera: Pyralidae) eggs treated with fungicides registered for tomato cultivation in selectivity tests using the methodology recommended by the IOBC/WPRS for selectivity studies. A total of 29 fungicides were evaluated in the experiments carried out at the Núcleo de Desenvolvimento Cientifico e Tecnologico em Manejo Fitossanitá (NUDEMAFI). Cards containing 20 eggs of the host were immersed in fungicide syrup and subjected to parasitism for 24 hours. After the parasitoid incubation period and their emergence, the effects of the fungicides on the biological characteristics of the parasitism were verified. It was concluded that the fungicides Supera® (COVER HYDROXIDE 537.4g/L), Nativo® (TEBUCONAZOLE 200g/L + TRIFLOXYSTROBIN 100g/L), Censor® (FENAMIDONE 500g/L), Manzate 800® (MANCOZEB 800g/Kg), Captan SC® (CAPTAN 480g/L), Amistar 500 WG® (AZOXYSTROBIN 500g/Kg), Frowncide 500 SC® (FLUAZINAM 500g/L), Fegatex® (BENZALCONIUM CHLORIDE 100g/L), Redshield 750® (CUPROSOXIDE 860g/Kg), Proplant® (PROPAMOCARB CHLORIDRATE 722g/L), Isatalonil 500 SC® (CHLOROTALONIL 750g/Kg), Recop® (COVER OXYCLORIDE 840g/Kg), Bion 500 WG® (ACIBENZOLAR-S-METHYL 500g/Kg) and Curzate BR® (CYMOXANIL 80g/Kg + MANCOZEB 640g/Kg) showed the best results for selectivity for the parasitoid *T. pretiosum. pretiosum.*

Keywords: Biological control. Phytosanitary products. Selectivity.

3.1 - INTRODUCTION

Chemical control is the most widely used method in conventional tomato farming, but although effective in most cases, pesticides can negatively affect the community of natural enemies, especially parasitoids. Biological and chemical controls need to be made compatible in order to preserve the community of beneficial insects in the production system. In this sense, the working group of the International Organization for Biological Control of Noxious Animals and Plants (IOBC), West Palaearctic Regional Section (WPRS) has established guidelines for conducting bioassays and categorizing pesticides in terms of selectivity, including the laboratory stages (adult and immature phases) (HASSAN et al., 2000; HASSAN; ABDELGADER, 2001).

Based on the principle of improving the efficiency of pest control, species of natural enemies have been used to manage insect pests in agricultural systems. The egg parasitoids of the genus Trichogramma are the ones that are being used on a commercial scale (PINTO, 2006) and allow for the optimization of lepidopteran pest control in the tomato crop (CAMPBELL et al., 1991). This association can reduce the number of pesticide applications, providing greater savings and less impact on the environment (LEBEDENCO et al., 2007).

These parasitoids are capable of parasitizing more than 200 species of insects that cause economic damage, mainly Lepidoptera (PRATISSOLI et al., 2004), in which they cause the death of developing embryos. However, in order to successfully use Trichogramma in crops, knowledge of the biological characteristics of the populations is fundamental (PRATISSOLI; PARRA, 2001).

Selectivity is fundamental to the phytosanitary management of pests in systems that aim to reduce the population of harmful insects, without altering or impacting the components of the ecosystem and the environment (SANTOS et al., 2006). In addition to the direct influence on pest eggs, fungicide-containing dressings can alter the behavior of the parasitoid, such as rejecting that potential host, thus preventing the desired parasitism from occurring.

In order to make biological and chemical control compatible, the aim of this study was to evaluate the biological characteristics of *Trichogramma pretiosum* when parasitized on eggs treated with fungicides, as well as to classify insecticides in terms of their impact on natural enemies, according to the recommendations of the IOBC/WPRS.

3.2 MATERIAL AND METHODS

The selectivity experiments were carried out in the Entomology sector of the Núcleo de Desenvolvimento Científico e Tecnológico em Manejo Fitossanitário (NUDEMAFI) of the Centro de Ciências Agrárias da Universidade Federal do Espirito Santo (CCA UFES), Alegre-ES. All the experiments were carried out in an air-conditioned chamber with the following characteristics: temperature 25 ± 1 °C, relative humidity (RH) 70 ± 10% and 14-hour photophase.

3.2.1 Breeding and maintenance of *Trichogramma pretiosum*

To maintain and multiply the parasitoid *T. pretiosum*, the alternative host *Anagasta kuehniella* (Zeller, 1879) (Lepidoptera: Pyralidae) was used, according to the methodology developed by Pratissoli et al. (2010a). To maintain the host, a diet based on whole wheat flour (70% m/m), corn flour (27% m/m) and brewer's yeast (3% m/m) was used.

To raise the parasitoid, eggs from the host were adhered to blue cardboard cards (8 x 2 cm) using gum arabic diluted to 50% (w/v). These eggs were then subjected to impregnation under a germicidal lamp for 50 minutes, as described by Pratissoli et al. (2010a), after which they were exposed to parasitism for a period of 24 hours and kept in an air-conditioned chamber at 25±1°C, RH of 70±10% and photophase of 14 hours.

3.2.2 - **Fungicides evaluated**

The products used in the test are shown in Table 3.

Table 3. List of fungicide active ingredients, concentrations of active ingredients and their respective doses used in the bioassays.

Technical name	**[ ] i.a.***	**Recommended application dose**	**Dose applied in 200 ml**
ACIBENZOLAR-S-METHYL	500 g/kg	5 g/100 L of water	0,01 g
AZOXYSTROBIN	500 g/kg	80 to 160 g/ha	0,026 g
CAPTAN 1	480 g/L	250 mL/100 L of water	0.5 mL
CAPTAN 2	500 g/kg	1.6 to 2.5 kg/ha	0,48 g
BENZALKONIUM CHLORIDE	100 g/L	250 mL/100 L of water	0.5 mL
PROPAMOCARB HYDROCHLORIDE 1	625 g/L	1.5 L/ha up to 300	0,416 g
PROPAMOCARB HYDROCHLORIDE 2	722 g/L	mL/100L of syrup	0.6 mL
CHLOROTHALONIL + DIMETHOMORPH	500 + 100 g/L	300 mL/100 L of water	0.6 mL
CHLOROTHALONYL	750 g/kg	200 g/100 L of water	0.6 mL
CYMOXANIL + MANCOZEB	80 + 640 g/kg	300 g/100 L of water	0,6 g
DIFENOCONAZOLE	250 g/L	up to 50 mL/100 L of water	0.1 mL
DIMETHOMORPH + MANCOZEB	90 + 600 g/kg	400 g/100 L of water	0,8 g
FENAMIDONE	500 g/L	300 mL/ha	0.06 mL
FLUAZINAM	500 g/L	100 mL/100 L of water	0.2 mL
COPPER HYDROXIDE 1	537.4 g/L	up to 4 L/ha	0.8 mL
COPPER HYDROXIDE 2	538 g/kg	up to 300 g/100 L of water	0,6 g
MANCOZEB + METALAXYL-M	640 + 40 g/kg	up to 2.5 kg/ha	0,6 g
MANCOZEB 1	800 g/kg	3 kg/ha	0,5 g
MANCOZEB 2	800 g/kg	up to 3 kg/ha	0,5 g
MANDIPROPAMID	250 g/L	0.4 á 0.6 L/ha	0.12 mL
METIRAM + PYRACLOSTROBIN	550 + 50 g/kg	400 g/100 L of water	0,8 g
METIRAM	700 g/kg	up to 3 kg/ha	0,6 g
COPPER OXYCHLORIDE	840 g/kg	up to 200 g/100 L of water	0,4 g
CUPROSOXIDE	860 g/kg	up to 150 g/100 L of water	0,48 g
PYRACLOSTROBIN	250 g/L	40 mL/100 L of water	0.08 mL
PROPINEB	700 g/kg	3 kg/ha	0,67 g
TEBUCONAZOLE + TRIFLOXYSTROBIN	200 + 100 g/L	up to 0.75 L/ha	0.2 mL
THIOPHANATE-METHYL 1	700 g/kg	70 g/100 L of water	0,14 g

ACIBENZOLAR-S-METHYL	500 g/kg	5 g/100 L of water	0,01 g
THIOPHANATE-METHYL 2	700 g/kg	up to 70 g/100 L of water	0,114 g

* Concentration of the active ingredient in the commercial product.

The choice of products was suggested taking into account the intensity of use by farmers in the producing region of Espirito Santo and registered with the Espirito Santo Agricultural and Forestry Defense Institute (IDAF).

Newly emerged females *of T. pretiosum* (age < 24 hours) were individualized in Eppendorf tubes (2.5 ml) containing a drop of honey inside for feeding. Sky blue cardboard cards (0.5 x 3.0 cm) containing 20 eggs of the host *A. kuehniella* glued together with gum arabic (25% m/v) were rendered unviable by exposure to a germicidal lamp for approximately 50 minutes.

The cartons containing the eggs were immersed in the product syrups for a period of five seconds. In the control treatment, only distilled water was used. The cards were then placed on paper towels in plastic trays in an exhaustion chamber for one hour at room temperature to remove excess moisture from the surface of the eggs. The cartons containing the treated eggs were offered to parasitized *T. pretiosum* for 24 hours. After this period, the females were removed and the cards kept in the same tubes for observation until the parasitoids emerged.

The parameters observed were the percentage of parasitism. Each treatment consisted of 15 replicates represented by a card containing 20 eggs of the host under study.

3.2.3 **Statistical analysis**

The design used was entirely randomized with 29 treatments (28 fungicides + control) and 15 replicates each. The parasitism data obtained in the bioassays was submitted to the parasitism reduction formula and then to analysis of variance (ANOVA) and the means were compared using the Scott-Knott test at 5% probability. The statistical analysis was carried out using the statistical program SISVAR version 5.1 (FERREIRA, 2003).

The reduction in parasitism of the parasitoid for each treatment was determined by comparison with the control (distilled water) and calculated using the formula: RP = (1 - Rt/Rc) x 100, where RP is the percentage reduction in parasitism or emergence; Rt is the average parasitism or emergence value for each product and Rc is the average parasitism or emergence observed for the control (negative) treatment (HASSAN et al., 2000). Based on these reduction percentages, the products tested were classified into classes, according to the IOBC/WPRS recommendations (1988 and 1992), where: 1 = harmless (< 30%), 2 = slightly harmful (30 - 79%), 3 = moderately harmful (80 - 99%) and 4 = harmful (> 99%).

3.2.4 RESULTS AND DISCUSSION

The action of fungicides can be influenced by the molecule, concentration or dose applied or by the biological organisms (parasitoid and host). It was observed that the parasitoid did not accept the host when they were immersed in some fungicide syrups. This may be due to contact with the antennae or even with the ovipositor itself and walking during the evaluation, which caused the potential host to be rejected, preventing oviposition or subsequently inhibiting the parasitoid's development (MOREIRA et al. 2009). Based on the parasitism data analyzed, statistical differences were observed between the active ingredients of the fungicides, which made it possible to subdivide them into five distinct groups and four classes of impact on natural enemies, according to the IOBC/WPRS recommendations.

The fungicides tested affected the parasitism of *T. pretiosum* as shown in Table 4. The fungicides THIOPHANATE METHYL 2, DIFENOCONAZOLE, COVER HYDROXIDE 2, CYMOXANIL + MANCOZEB, MANCOZEB + METALAXYL-M, PROPAMOCARB CHLORIDRATE 2, METIRAM, ACIBENZOLAR-s-METHYL, COVER OXYCHLORIDE and CHLOROTHALONIL were grouped together because they showed no difference in terms of average parasitism, with a range from 85.77 to 72.08%, and promoted an increase in the reduction of parasitism from 36.72 to 14.90% to which they were classified as innocuous according to the IOBC/WPRs classification (Table 4).

Table 4. Active ingredients of fungicides organized according to the average percentage of parasitism observed, reduction in parasitism of *T. pretiosum* in relation to the control and IOBC/WPRS classes.

Active ingredients of fungicides	**Parasitism (%)***	**PR (%)**[1]	**Class**[2]
THIOPHANATE METHYL 2	85,77 a	(36,72) **	1
DIFENOCONAZOLE	85,66 a	(36,55) **	1
COPPER HYDROXIDE 2	85,41 a	(36,16) **	1
CYMOXANIL + MANCOZEB	82,86 a	(32,08) **	1
MANCOZEB + METALAXYL-M	81,25 a	(29,51) **	1
PROPAMOCARB HYDROCHLORIDE 2	78,89 a	(25,75) **	1
METIRAM	78,57 a	(25,24) **	1
ACIBENZOLAR-S-METHYL	75,00 a	(19,55) **	1
COPPER OXYCHLORIDE	73,33 a	(16,89) **	1
CHLOROTHALONYL	72,08 a	(14,90) **	1
PROPAMOCARB HYDROCHLORIDE 1	69,64 b	(11,01) **	1
CUPROSOXIDE	69,23 b	(10,36) **	1
BENZALKONIUM CHLORIDE	68,57 b	(9,30) **	1
FLUAZINAM	66,66 b	(6,26) **	1
CHLOROTHALONIL + DIMETHOMORPH	64,38 b	(2,61) **	1
AZOXYSTROBIN	62,50 b	0,37	1
CAPTAN 2	61,33 b	2,23	1
MANCOZEB 2	60,41 b	3,69	1

FENAMIDONE	57,14 b	8,91	1
TEBUCONAZOLE + TRIFLOXYSTROBIN	55,41 b	11,66	1
COPPER HYDROXIDE 1	44,09 c	29,71	1
PROPINEB	40,71 c	35,09	2
PYRACLOSTROBIN	12,27 d	80,43	3
CAPTAN 1	e	100,00	4
DIMETHOMORPH + MANCOZEB	e	100,00	4
MANCOZEB 1	e	100,00	4
METIRAM + PYRACLOSTROBIN	e	100,00	4
THIOPHANATE METHYL 1	e	100,00	4
WITNESS	62,73 b		
CV (%)	25,19		
F	11,96		
GLR	279		
P	<0,001		

* Averages followed by the same letter do not differ statistically by the Scott-Knott test at 5% probability;

** Parasitism Reduction (PR) values in brackets indicate an increase in parasitism compared to the control;

[1] PR = Reduction in parasitism compared to the control.

[2] IOBC/WPRS classes for selectivity testing on Trichogramma: 1 = innocuous (< 30%), 2 = slightly harmful (30-79%), 3 = moderately harmful (80-99%), 4 = harmful (> 99%).

Although they differed from the previous group, the fungicides PROPAMOCARB 1 CHLORIDRATE, CUPROSOXIDE, BENZALCONIUM CHLORIDE, FLUAZINAM, CHLOROTHALONIL + DIMETHOMORPH, AZOXYSTROBIN, CAPTAN 2, MANCOZEB 2, FENAMIDONE and TEBUCONAZOLE + TRIFLOXYSTROBIN were grouped as intermediate fungicides in terms of the parasitism observed in the tests, ranging from 69.64 to 55.41% and parasitism reduction values from (11.01) to 11.66% where the groups, although distinct, did not differ in terms of the classification recommended by the IOBC/WPRS, both being classified as innocuous (class 1). In the same way as before, the fungicide COVER HYDROXIDE 1 differed from the previous groups with average parasitism values of 44.09% and a reduction in parasitism of 29.71%, being classified as innocuous according to the IOBC/WPRS classification.

The fungicide with the active ingredient PROPINEB was the only one to be classified as slightly harmful to the parasitoid, with an average parasitism of 40.71% and a reduction in parasitism of 35.09%, being classified as class 2 according to the IOBC/WPRS classification. The fungicide based on PYRACLOSTROBIN was the only representative in the IOBC/WPRS classification as moderately harmful to the parasitoid, with an average parasitism value of 12.27% and a reduction in parasitism of 80.43%, being classified as class 3 according to the IOBC/WPRS classification.

For the fungicides with the active ingredient CAPTAN 1, DIMETHOMORPH + MANCOZEB,

MANCOZEB 1, METIRAM + PIRACLOSTROBIN and THIOPHANATE METHYL 1, no parasitism was observed by the parasitoid *T. pretiosum. pretiosum*, a fact which implies that parasitism was not observed on the eggs of the host. These reasons may be related to the influence of the molecule, concentration or dose applied to the fungicides or even the biological organisms involved in the bioassays (parasitoid and host). For the purposes of the IOBC/WPRS classification, they were classified as harmful (class 4) to the egg parasitoid *T. pretiosum*.

Extreme parasitism values were observed for fungicides based on THIOPHANATE METHYL, even though these products had the same concentration of the active ingredient. Host selection by Trichogramma is carried out by means of chemical and physical stimuli received by the antennae (COLAZZA et al., 2010). Therefore, these results may be directly related to the composition of the inerts in these products, since at the time of the host selection analysis by Trichogramma, some constituent must have induced the parasitoid to reject.

Other researchers working with fungicides based on THIOPHANATE METHYL reported favorable results for the use of this active ingredient associated with *T. pretiosum* on *A. kuehniella* eggs (MANZONI et al., 2006; CARVALHO et al., 2012). On the other hand, on *Trichogramma atopovirilia* Oatman & Platner (Hymenoptera: Trichogrammatidae) in eggs of *Diaphania hyalinata* Linnaeus (Lepidoptera: Pyralidae), the fungicide THIOPHANATE METHYL reduced parasitism by 27.64% (PRATISSOLI, et al., 2010).

The fungicides based on copper hydroxide also showed different results on *T. pretiosum* parasitism, despite the proximity of the concentrations of the active ingredients. However, in relation to the IOBC/WPRS classification, these fungicides were the same and fell into class 1. On *A. kuehniella* eggs, Carvalho et al. (2012) reported a reduction in the parasitism rate of 80% for *T. pretiosum*, which classified COBURN HYDROXIDE as class 3.

The difference observed in *T. pretiosum* parasitism in eggs treated with CAPTAN-based fungicides could possibly be associated with the formulation of the products, since CAPTAN 1 is a concentrated suspension formulation while CAPTAN 2 is a wettable powder, or with some characteristic of the selected parasitoid, such as age. Manzoni et al. (2006) in their study found no difference for the CAPTAN fungicide in relation to the formulation, with high parasitism values for *T. pretiosum* on *A. kuehniella* eggs. Thus, the result obtained in this study is related to the age of the Trichogramma, since their emergence is not homogeneous.

The lack of parasitism by *T. pretiosum* observed for the treatments DIMETHOMORPH + MANCOZEB, MANCOZEB 1 and METIRAM + PIRACLOSTROBIN show that these products are extremely toxic to the parasitoid. As with the CAPTAN-based fungicide, the MANCOZEB-based fungicides showed different results, especially the single ones. For solitary MANCOZEB, Pratissoli et al. (2010) reported values of over 60% of parasitism *of T. atopovirilia* in *D. hyalinata* eggs, while Carvalho et al. (2012) reported that the association of MANCOZEB with other active ingredients can intensify the toxicity of fungicides, such as those observed for the association between DIMETHOMORPH + MANCOZEB. No toxicity levels were observed for the combination of CYMOXANIL + MANCOZEB and MANCOZEB + METALAXYL-M.

According to some authors, the increase in the parasitism rate is related to the possible occurrence of hormesis (ZANUNCIO et al., 2003; PRATISSOLI et al., 2010; CARVALHO et al., 2012). Hormesis is based on the provision of favorable biological responses to beneficial organisms, when exposed to stressors in small quantities, as stimulatory effects (CALABRESE, 2010; GUEDES; MAGALHAES; COSME, 2009; MATTSON; CALABRESE, 2010). However, confirmation of this event requires detailed studies on chemical analysis for more complex biological organisms, such as endoparasitoids.

3.4 CONCLUSION

Among the products tested, fungicides with the active ingredient THIOPHANATE METHYL 2, DIFENOCONAZOLE, COBRE HYDROXIDE 2, CYMOXANIL + MANCOZEB, MANCOZEB + METALAXYL-M, PROPAMOCARB 2 CHLORIDRATE, METIRAM, ACIBENZOLAR-S-METHYL, COBRE OXYCHLORIDE, CHLOROTHALONYL, PROPAMOCARB 1 CHLORIDRATE, CUPROSOXIDE, BENZALCHONIUM CHLORIDE, FLUAZINAM and CHLOROTHALONYL + DIMETHOMORPH, as well as being used as disease control agents, promoted an increase in the parasitism of the parasitoid *T. pretiosum when* applied to the eggs of the *plant. pretiosum* when applied to eggs of the host *A. kuehniella*. These active ingredients of fungicides should be prioritized when necessary in disease control because they make biological control, using *T. pretiosum*, and chemical control compatible.

3.5 REFERENCES

ALMEIDA, G. D.; ZANUNCIO, J. C.; PRATISSOLI, D.; ANDRADE, G. S.; CECON, P. R.; SERRAO, J. E. Effect of azadirachtin on the control of *Anticarsia gemmatalisand* its impact on *Trichogramma pretiosum*. **Phytoparasitica**, v. 38, p. 413-419, 2010.

ALVES, F.R.; JESUS JUNIOR, W.C.; PRATISSOLI, D.; POLANCZYK, R.A.; ZANUNCIO JUNIOR, J.S.; HOLTZ, A.M.; VIANNA, U.R. Phytosanitary management of diseases and pests: new perspectives. In: JESUS JUNIOR, W.C.; POLANCZYK, R.A.; PRATISSOLI, D.; PEZZOPANE, J.E.M.; SANTIAGO, T. (Eds.). **Updates on plant health defense**. Alegre, Federal University of Espirito Santo, ch. 13, p. 383-416, 2007.

BESTETE, L.R; PRATISSOLI, D.; QUEIROZ, V.T. de; CELESTINO, F.N; MACHADO, L.C Toxicity of castor oil to Helicoverpa zea and Trichogramma pretiosum. **Pesquisa Agropecuaria Brasileira**, Brasilia, v.46, n.8, p.791-797, 2011.

CARMO, E.L. do; BUENO, A.F.; BUENO, R.C.O.F.; VIEIRA, S.S.; GOULART, M.M.P.; CARNEIRO, T.R. Selectivity of phytosanitary products used in soybean cultivation for pupae *of Trichogramma pretiosum* Riley, 1879 (Hymenoptera: Trichogrammatidae). **Arquivos do Instituto Biológico**, v. 77, n. 2, p. 283-290, 2010.

CARVALHO, G. A.; GODOY, M. S.; PARREIRA, D. S.; REZENDE, D. T. Effect of chemical insecticides used in tomato crops on immature Trichogramma pretiosum (Hymenoptera: Trichogrammatidae). **Revista Colombiana de Entomologia** 36 (1): 10-15 (2010).

CARVALHO, G. A.; PARRA, J. R. P.; BAPTISTA, G. C. Bioactivity of phytosanitary products used in tomato cultivation (*Lycopersicon esculentum* MILL.) to *Trichogramma pretiosum* Riley, 1879 (Hymenoptera: Trichogrammatidae) in F1 and F2 generations. **Ciência e Agrotecnologia**, v. 27, n. 2, p. 261-270, 2003.

CARVALHO, J. R.; PRATISSOLI, D.; PAES, J. P. P.; ZUIM, V.; STINGUEL, P. SELECTIVITY OF FUNGICIDES USED IN THE CULTURE OF TOMATOES (*Lycopersicum esculentum*, MILL.) TO *Trichogramma pretiosum*. **Nucleus** (Ituverava. Online), v. 9, p. 1-8, 2012.

COLAZZA, S.; PERI, E.; SALERNO, G.; CONTI, E. Host Searching by Egg Parasitoids: Exploitation of Host Chemical Cues. Chap 4, p. 97-147. In: CÔNSOLI F.L., PARRA J.R.P., ZUCCHI R. (eds.), **Egg Parasitoids in Agroecosystems with Emphasis on Trichogramma**, Progress in Biological Control 9, Springer, Piracicaba, 2010.

CZEPAK, C.; FERNANDES, P.M.; ALBERNAZ, K.C.; RODRIGUES, O.D.; SILVA, L.M; SILVA, E.A.; TAKATSUKA, F.T.; BORGES, J.D. Selectivity of insecticides to the complex of natural enemies in the cotton crop (Gossypium hirsutum).

Pesquisa Agropecuâria Tropical, v.35, n.2, p.123-127, 2005.

FERREIRA, D.F. **SISVAR**: version 5.1. Lavras: UFLA, 2003.

GIOLO, F. P.; GRÜTZMACHER, A. D.; MANZONI, C. G.; LIMA, C. A. B. NORNBERG, S. D. Toxicity of phytosanitary products used in peach cultivation on adults of *Trichogramma pretiosum*. **Bragantia**, v. 66, p.423-431, 2007.

GOULART, R.M.; BORTOLI, S.A.; THULER, de, R.T.; D. PRATISSOLI; VIANA, C.L.T.P.; H.X.L. VOLPE. Evaluation of the selectivity of insecticides to *trichogramma* spp. (hymenoptera: trichogrammatidae) in different hosts. **Archive: Instituto Biologico de Sao Paulo**, Sao Paulo, v.75, n.1, p.69-77, jan./mar., 2008.

HASSAN, S.A.; HAFES, H.; DEGRANDE, P.E.; HERAI, H. The side effects of pesticides on the egg parasitoid *Trichogramma caecoeciae* Marchal (Hym., Trichogrammatidae), acute dose-response and persistence tests. **Journal of Applied Entomology**, v. 122, n. 9-10, p. 569-573, 1998.

HASSAN, S.A.; HALSALL, N.; GRAY, A.P.; KUEHNER, C.; MOLL, M.; BAKKER, F.M.; ROEMBKE, J.; YOUSEF, A.; NASR, F.; ABDELGADER, H. A laboratory method to evaluate the side effects of plant protection products on *Trichogramma cacoeciae* Marchal (Hym., Trichogrammatidae). In: CANDOLFI, M.P.; BLÜMEL, S.; FORSTER, R.; BAKKER, F.M.; GRIMM, C.; HASSAN, S.A.; HEIMBACH, U.; MEAD- BRIGGS, M.A.; REBER, B.; SCHMUCK, R.; VOGT, H. (Eds.). **Guidelines to evaluate side-effects of plant protection products to non-target arthropods.** IOBC/ WPRS, Gent. p. 107-119, 2000.

LUZ, J. M.Q.; SHINZATO, A. V.; SILVA, M. A. D. Comparison of conventional and organic tomato production systems in protected cultivation. **Bioscience Journal**, v. 23, p. 7-15, 2007.

MOURA, A. P.; CARVALHO, RIGITANO, G. A.; R. L. de O. Toxicity of insecticides used in tomato cultivation to *Trichogramma pretiosum.* **Pesquisa Agropecuaria Brasileira**, Brasilia, v.40, n.3, p.203-210, mar. 2θO5.

PRATISSOLI, D.; DALVI, L. P.; POLANCZYK, R. A.; ANDRADE, G. S.; HOLTZ, A. M.; NICOLINE, H. O. Biological characteristics of *Trichogramma exiguum* on eggs of *Anagasta kuehniella* and *Sitotroga cerealella*. **Idésia**, v. 28, n. 1, p. 39-42, 2010a.

PRATISSOLI, D.; FORNAZIER, M.J.; HOLTZ, A.M.; GONÇALVES, J.R.; CHIORAMITAL, A.B.; ZAGO, H. Occurrence of *Trichogramma pretiosum* in commercial tomato areas in Espirito Santo, in regions of different altitudes. **Horticultura Brasileira,** Brasilia, v. 21, n. 1, p. 73-76, 2002.

PRATISSOLI, D.; MILANEZ, A. M.; BARBOSA, W. F.; CELESTINO, F. N.; ANDRADE, G. S.; POLANCZYK, R. A. SIDE EFFECTS OF FUNGICIDES USED IN CUCURBITACEOUS CROP ON *Trichogramma atopovirilia* OATMAN & PLATNER (HYMENOPTERA: TRICHOGRAMATIDAE). **Chilean Journal of Agricultural Research**, v. 70, n. 2, p. 323-

327, 2010b.

PRATISSOLI, D.; MILANEZ, A. M.; CELESTINO, F. N.; BARBOSA, W. F.; VIANNA, U. R.; POLANCZYK, R. A.; ZINGER, F. D.; CARVALHO J. R. de. Selectivity of insecticides recommended for cucurbits for *Trichogramma atopovirilia* Oatman & Platner (Hymenoptera: Trichogrammatidae) under laboratory conditions. **Rev. Ceres**, Viçosa, v. 58, n.5, p. 661-664, Sep/Oct, 2011.

PRATISSOLI, D.; VIANNA, U. R.; FURTADO, G. O.; ZANUNCIO, J. C.; POLANCZYK, R. A.; BARBOSA, W. F.; CARVALHO, J. R. Selectivity of insecticides to *Trichogramma pretiosum* Riley (Hymenoptera: Trichogrammatidae) in different hosts. **Boletin de Sanidad Vegetal Plagas**, v. 35, n. 3, p. 347-353, 2009.

RUSSELL, J. E.; STOUTHAMER, R. Sex Ratio Modulators of Egg Parasitoids. In: CÔNSOLI, F. L.; PARRA, J. R. P.; ZUCCHI, R. (Eds.) Egg Parasitoids in Agroecosystems with Emphasis on *Trichogramma*. **Springer**, Piracicaba, p. 167-190, 2010.

SCHOLZ B.C.G., ZALUCKI M.P. 2000. The effects of two new insecticides on the survival of adult Trichogramma pretiosum Riley in sweet corn. In: Austin A, Dowton M, editors. **Hymenoptera: Evolution, Biodiversity and Biological Control.** Canberra, Australia: CSIRO Pubs. p. 381-387.

SOUZA, J.C.; REIS, P.R. Main pests of table tomatoes: bioecology, damage and control. **Informe Agropecuàrio**, v.24, p.79-92, 2003.

TROUGAKOS, I., L. MARGARITIS. 2002. Novel morphological and physiological aspects of insect eggs. p. 3-36. In M. Hilker and T. Meiners (eds.). Chemoecology of insect eggs and egg deposition. **Blackwell Publishing Ltd.**, Berlin, Germany.

VINSON, S. B. 1997. Host selection behavior of egg parasitoids, with emphasis on the Trichogrammatidae family, 67-120. In: PARRA, J. R. P., ZUCCHI, R. A. (eds.). *Trichogramma* and applied biological control. Piracicaba: FEALQ, 324pp. WAJNBERG, E. Genetics of the Behavioral Ecology of Egg Parasitoids. In: CÔNSOLI, F. L.; PARRA, J. R. P.; ZUCCHI, R. (Eds.) Egg Parasitoids in Agroecosystems with Emphasis on *Trichogramma*. **Springer**, Piracicaba, p. 149-166, 2010.

ZANUNCIO, T.V., SERRÂO, J. E.; ZANUNCIO, J. C.; GUEDES, R. N. C. Permethrin-induced hormesis on the predator *Supputius cincticeps* (Stal, 1860) (Heteroptera: Pentatomidae). **Crop Protection**. v. 22, p. 941-947, 2003.

CHAPTER IV

CAN THE HOSPITAL INFLUENCE THE SELECTIVITY OF FUNGICIDES AND INSECTICIDES FOR *Trichogramma pretiosum?*

SUMMARY

The aim of this study was to evaluate the influence of different insecticides and fungicides on *Trichogramma pretiosum* (Riley, 1879) (Hymenoptera: Trichogrammatidae) in different hosts. The tests were carried out on adults of *T. pretiosum* on the hosts *Anagasta kuehniella* (Zeller, 1879) (Lepidoptera: Pyralidae) and *Helicoverpa zea* (Boddie, 1850) (Lepidoptera: Noctuidae). The results showed that the pesticides studied had different effects on *T. pretiosum* depending on the host. The insecticides IMIDACLORIPRID + BETA-CIFLUTRIN and METHOMYL were the most aggressive, reducing the parasitism *of T. pretiosum* adults on both hosts, while for the fungicides, DIMETOMORPH + MANCOZEB and MANCOZEB were the most influential. When treated with fungicides, there was a reduction in emergence only for DIMETOMORPH + MANCOZEB. In view of the results, it can be concluded that the host can influence the selective action of insecticides and fungicides on parasitoids of the Trichogramma genus. Furthermore, *T. pretiosum* was more susceptible to the insecticide METHOMYL and ACIBENZOLAR-S-METHYL when reared on *A. kuehniella* and to the fungicide DIMETOMORPH + MANCOZEB when reared on *H. zea*.

Keywords: Parasitoid. Pesticides. Phytosanitary management.

4.1 INTRODUCTION

The quest to reduce losses in tomato plantations has led farmers to make massive and indiscriminate use of pesticides due to phytosanitary problems (CARVALHO et al., 2003; CARVALHO et al., 2012). Among these, the crop's high susceptibility to diseases and the frequent presence of insect pests are the main obstacles in the production chain, which classifies tomatoes as a high-risk crop (SOUZA; REIS, 2003; LUZ et al., 2007).

In order to reduce the use of these products, *primarily* insecticides, other methods of pest management should be used or combined with pesticides to reduce insect pests (ALVES et al., 2007). One management method that has been widely studied and used is biological control using parasitoids of the Trichogramma genus (ALMEIDA et al., 2010). The use of these agents in association with pesticides can be interfered with due to the action of these products. In view of this, it is necessary to carry out tests to verify the influence of pesticides on these organisms, generating information that can help in decision-making in

Phytosanitary Pest Management programs (PRATISSOLI et al., 2002; ZANUNCIO et al., 2003; CARVALHO, et al., 2012).

The protocol for standardizing selectivity and susceptibility tests to assess the toxicity and impact of pesticides on Trichogramma was developed by the International Organization for Biological Control of Noxious Plants and Animals/ West Palaearctic Region Section (IOBC/ WPRS), a working group that standardizes selectivity tests (HASSAN et al., 1998; PRATISSOLI et al., 2009).

Most selectivity tests are carried out on alternative hosts, such as *A. kuehniella*. The aim of this study was to evaluate the influence of insecticides and fungicides belonging to each toxicological class on *T. pretiosum* Riley in different hosts.

4.2 MATERIAL AND METHODS

The selectivity and susceptibility experiments were carried out in the Entomology sector of the Núcleo de Desenvolvimento Cientifico e Tecnolòico em Manejo Fitossanitàrio (NUDEMAFI) of the Centro de Ciências Agràrias da Universidade Federal do Espirito Santo (CCA-UFES), in Alegre - ES. All the experiments were carried out in an Eletrolab model 102FC air-conditioned chamber, regulated with the following characteristics: temperature 25 ± 1 °C, relative humidity (RH) 70 ± 10% and 14-hour photophase.

4.2.1 Breeding and maintenance of *Trichogramma pretiosum*

To maintain and multiply the parasitoid *T. pretiosum*, the alternative host *A. kuehniella* was used, according to the methodology developed by Pratissoli et al. (2010a). To maintain the host, a diet based on whole wheat flour (70% m/m), corn flour (27% m/m) and brewer's yeast (3% m/m) was used.

To raise the parasitoid, eggs from the host were adhered to blue cardboard cards (8 x 2 cm) using gum arabic diluted to 50% (w/v). These eggs were then subjected to impregnation under a germicidal lamp for 50 minutes, as described by Pratissoli et al. (2010a), after which they were exposed to parasitism for a period of 24 hours and kept in an air-conditioned chamber at 25±1°C, 70±10% RH and 14-hour photophase.

4.2.2 Pesticides tested

The products used in the tests are shown in Table 5. The products were chosen by random drawing, taking into account only the separation of the products by class of use (insecticide or fungicide).

The selective action studies were divided into two stages, the first evaluating products belonging to the insecticide class and the second evaluating products belonging to the fungicide class.

Table 5. Pesticides used in selectivity and susceptibility tests on Trichogramma.

Technical name	**[]i.a.***	**Formula**	**Classification TOXICOLOGY**	**Manufacturer**
	Experiment I - Insecticides			
IMIDACLOPRID + BETA-CYFLUTHRIN	100 + 12.5 g/l	SC	II - Highly toxic	Bayer CropScience
METHoMYL	215 g/l	SL	I - Extremely toxic	Du Pont do Brasil S/A
ACETAMiPRiD	200 g/kg	WP	III - Medium toxic	Iharabras
SPiNoSAD	480 g/l	SC	IV - Slightly toxic	Dow AgroSciences
	Experiment II - Fungicides			
DiMEToMoRPH + MANCoZEB	90 + 600 g/kg	WP	II - Highly toxic	Basf
ACiBENZoLAR-S-METHYL	500 g/kg	WG	III - Medium toxic	Syngenta
MANCoZEB + oXiCLoRETo DE CoBRE	440 + 300 g/kg	WP	IV - Slightly toxic	Sipcam
MANCoZEB	800 g/kg	WP	I - Extremely toxic	Dow AgroSciences

* Concentration of the active ingredient in the commercial product.

The selectivity/susceptibility studies were divided into two experiments. Experiment I evaluated products belonging to the insecticide class, while Experiment II evaluated products belonging to the fungicide class.

4.2.3 Selectivity bioassay

Newly emerged females *of T. pretiosum* (age < 24 hours) were individualized in Eppendorf tubes (2.5 ml) containing a drop of honey inside for feeding. Sky blue cardboard cards (0.5 x 3.0 cm) containing 20 eggs of the host *A. kuehniella* or the host *H. zea* glued together with gum arabic (25% m/v) were rendered unviable by exposure to a germicidal lamp for approximately 50 minutes.

The cards containing the eggs were immersed in the syrups of the products belonging to Experiments I or II for a period of five seconds. In the control treatment, only distilled water was used as a standard. The cartons were then placed on paper towels in plastic trays in an exhaustion chamber for one hour at room temperature to remove excess moisture from the surface of the eggs. The cartons containing the treated eggs were offered to parasitized *T. pretiosum* for 24 hours. After this period, the females were removed and the cards kept in the same tubes for observation until the parasitoids emerged.

The parameters observed were: percentage of parasitism and emergence. The number of individuals per egg was not evaluated because there was a difference in the size of the eggs of the hosts under study. Each treatment consisted of 15 replicates represented by a card containing 20 eggs of the host under study.

4.2.5 Statistical analysis

The design used for the selectivity bioassay was entirely randomized in a 5 x 2 factorial scheme (treatments x hosts) with 15 replicates each for each experiment.

The design used for the susceptibility bioassay was entirely randomized in a 5 x 2 factorial scheme (treatments x hosts) with 15 replicates each for each experiment and parasitoid development stage.

Statistical analysis was carried out using the statistical program SISVAR version 5.1 (FERREIRA, 2003). The data obtained was submitted to analysis of variance (ANOVA) and the means were compared using the Scott-Knott test at 5% probability.

The reduction in parasitism or parasitoid emergence for each treatment was determined by comparison with the control (distilled water) and calculated using the formula: RP = (1 - Rt/Rc) x 100, where RP is the percentage reduction in parasitism or emergence; Rt is the average parasitism or emergence value (a) for each product and Rc is the average parasitism or emergence observed (a) for the (negative) control treatment (HASSAN et al., 2000). Based on these reduction percentages, the products tested were classified into classes, according to the IOBC/WPRS recommendations (1988 and 1992), where: 1 = harmless (< 30%), 2 = slightly harmful (30 - 79%), 3 = moderately harmful (80 - 99%) and 4 = harmful (> 99%).

4.3 Results and discussion

4.3.1 Selectivity

Analysis of the parasitism data from Experiment I (insecticides) showed a significant interaction between the Treatment (insecticides) and Host factors (F=3.111; glr=107; p=0.0294) (Table 6). When unfolding the levels of the Insecticides factor within the levels of the Host factor, it was found that the parasitism of *T. pretiosum* was similar for the two hosts. The insecticide SPINOSAD was the one that least reduced the parasitism of *T. pretiosum* for both hosts (78.57 and 85.67% parasitism for the hosts *A. kuehniella* and *H. zea*, respectively). The insecticide IMIDACLOPRIDO+BETA-CYFLUTRIN was the one that most reduced parasitism (parasitism < 30%). When unfolding the levels of the Host factor within the levels of the Insecticides factor, there was a significant difference only for the treatments IMIDACLOPRIDO+BETA-CIFLUTRINA and ACETAMIPRID, with a greater reduction in parasitism on *H. zea* for the second insecticide.

Table 6. Percentage of parasitism, reduction in parasitism (RP) of *T. pretiosum* in relation to the control and selectivity classes of insecticides (Experiment I) recommended for tomato cultivation and percentage emergence of the parasitoid's offspring.

Parasitism (%) *								
Insecticides	Hosts							
	A. kuehniella		PR[1]	Class[2]	*H. zea*		PR[1]	Class[2]
IMIDACLOPRID + BETA-CIFLUTRIN	13,33	dB	83,81	3	26,25	cA	67,48	2
METHOMYL	47,92	cA	41,80	2	41,50	bA	48,58	2
ACETAMIPRID	62,50	bA	24,09	1	46,43	bB	42,48	2
SPINOSAD	78,57	aA	4,57	1	85,67	aA	(6,14)	---
WITNESS	82,33	aA	---		80,71	aA	---	
CV (%)					22,57			
F glr p				3,111 101 0,0294				

Emergence (%)*		
Insecticides	Hosts	
	A. kuehniella	*H. zea*
IMIDACLOPRID + beta-CIFLUTRIN	91.67 aA	100.00 aA
METHoMYL	30.16 bB	98.33 aA
ACETAMIPRID	89.41 aA	74.58 bA
SPiNoSAD	99.35 aA	100.00 aA
WITNESS	97.10 aA	98.71 aA
CV (%)	20,75	
F	20,947	
glr	101	
p	<0,001	

* Averages followed by the same letter, lower case in the column and upper case in the row, do not differ according to the Scott-Knott test ($p<0.05$);

[1] PR = Reduction in parasitism compared to the WITNESS. Value in brackets indicates an increase in parasitism;

[2] ioBC/WPRS classes for selectivity testing on *Trichogramma*: 1 = innocuous (< 30%), 2 = slightly harmful (30-79%), 3 = moderately harmful (80-99%), 4 = harmful (> 99%).

With regard to the IOBC classification, only the SPINOSAD treatment fell into class 1 (innocuous) for both hosts, and for the *H. zea* host the parasitism was higher than that of the specific control (85.67 and 80.71%, respectively) (Table 6). The METHOMYL treatment scored IOBC class 2 for both hosts. The ACETAMIPRID treatment was class 1 for the *A. kuehniella* host and class 2 for the *H. zea* host, while the IMIDACLOPRIDO+BETA-CIFLUTRINA treatment was class 3 and 2 for the *A. kuehniella* and *H. zea* hosts, respectively (Table 6).

The negative effect of METHOMYL on *T. pretiosum* was reported by Scholz and Zalucki (2000) on the host *Helicoverpa armigera* (Hübner, 1805) (Lepidoptera: Noctuidae), who in turn reported that this product is highly toxic to Trichogramma, sharply reducing its survival, which possibly led to the reduced parasitism rate observed in this study. Evaluating IMIDACLOPRID and BETA-CIFLUTRIN solitaries, Czepak et al. (2005) reported a reduction of more than 60% in the presence of natural enemies 3 and 7 days after application, a fact that can be explained by the high toxicity of these products (Toxicological Class II - Highly toxic) (Table 5).

For the parameter percentage emergence of *T. pretiosum* offspring, there was an interaction between the factors Insecticides and Host (F=20.947; glr=107; p<0.001) (Table 6). In general, the percentage of emergence was higher than 89%, except for the combinations of METHOMYL x host *A. kuehniella* and ACETAMIPRID x host *H. zea*, which were lower (30.16 and 74.58%, respectively).

Based on the results observed for METHOMYL, which is an extremely toxic product (Table 5), it can be seen that the host influenced the viability of parasitism (emergence) of *T. pretiosum*. This influence could possibly be related to the physico-chemical characteristics of the egg. Trougakos and Margaritis (2002) point out that the physico-chemical characteristics of insect eggs are peculiar to each species, since each insect has certain basic needs for survival in its specific habitat. In this way, the more heterogeneous the environment, the greater the physical and chemical structure of the eggs to guarantee the survival of the offspring.

For Experiment II (fungicides), there was no interaction between the levels of the Fungicide and Host factors for the parasitism parameter (F=0.115; glr=101; p=0.9515), nor was there an effect of the Host factor for this parameter (F=0.049; glr=101; p=0.8248). On the other hand, there was an effect of the Fungicide factor on the parasitoid (F=104.463; glr=101; p<0.001). All the fungicides negatively affected the parasitism *of T. pretiosum* (Table 7). The fungicide ACIBENZOLAR-S-METHYL caused the least reduction in parasitism, while the fungicide DIMETOMORPH+MANCOZEB had the greatest influence (75.00 and 22.65%, respectively).

Table 7. Percentage of parasitism, reduction in parasitism (RP) of *T. pretiosum* in relation to the control and selectivity classes of fungicides (Experiment II) recommended for tomato cultivation and percentage emergence of the parasitoid's offspring.

Fungicides	**Parasitism (%)***		**PR[1]**	**Class[2]**
DIMETOMORPH + MANCOZEB	22,65	e	73,06	2
ACIBENZOLAR-S-METHYL	75,00	b	10,78	1
MANCoZEB + COBRE OXYCLECT	43,33	c	48,45	2
MANCoZEB	32,35	d	61,51	2
WITNESS A. KUEHNIELLA	82,33 a		---	
WITNESS H. ZEA	85,80 a			
WITNESS (AVERAGE)	84,07			
CV (%)	19,79			
F	104,463			
glr	101			

p	<0,001			

Emergence (%)**

Fungicides	Hosts			
	A. kuehniella		*H. zea*	
DIMETOMORPH + MANCOZEB	100,00	aA	91,67	bB
ACIBENZOLAR-S-METHYL	100,00	aA	97,68	aB
MANCoZEB + COBRE oXICLECT	100,00	aA	100,00	aA
MANCoZEB	92,26	bB	98,44	aA
WITNESS	97,10	aA	98,71	aA
CV (%)	5,49			
F	5,571			
glr	101			
p	0,0014			

1 Means followed by the same lowercase letter in the column do not differ according to the Scott-Knott test ($p<0.05$);
2 * Averages followed by the same letter, lower case in the column and upper case in the row, do not differ according to the Scott-Knott test ($p<0.05$);
[1] PR = Reduction in parasitism compared to the control;
[2] IoBC/WPRS classes for selectivity testing on Trichogramma: 1 = innocuous (< 30%), 2 = slightly harmful (30-79%), 3 = moderately harmful (80-99%), 4 = harmful (> 99%).

Based on the IOBC classification, the treatments DIMETOMORPH + MANCOZEB, MACOZEB + COBRE OXYCLORIDE and MANCOZEB fell into class 2 (slightly harmful). The treatment ACIBENZOLAR-S-METHYL, on the other hand, was in class 1 (harmless) (Table 7).

The reduction in Trichogramma performance caused by MANCOZEB has already been reported (PRATISSOLI et al., 2010b; CARVALHO et al., 2012). The difference observed for the fungicides composed of MANCOZEB may be associated with the concentration of the active ingredient in the product, since they differ in terms of toxicological class (Table 5). Despite the lower concentration of MANCOZEB in the fungicide DIMETOMORPH + MANCOZEB compared to the fungicide MANCOZEB, the negative action was intensified when combined with DIMETOMORPH. The 48.45% reduction in parasitism observed for MACOZEB and COVER OXYCLORIDE in combination is higher than that reported by Giolo et al. (2007), who did not observe any reduction in *T. pretiosum* parasitism on *A. kuehniella* eggs. This difference may be associated with intrinsic characteristics of the strains under study, since different populations may have a greater or lesser degree of adaptation to stressors such as pesticides.

With regard to the percentage emergence paramotor, there was an interaction between the Fungicide and Host factors ($F=5.571$; $glr=101$; $p=0.0014$), ranging from 92.26 to 100% emergence (Table 7). Analyzing the breakdown of the Fungicides factor within the Host factor levels, the MANCOZEB fungicide showed the lowest percentage of emergence of *T. pretiosum* offspring for the *A. kuehniella* host (92.26%), while for the *H. zea* host the lowest emergence occurred for the DIMETOMORPH + MANCOZEB treatment (91.67%). When analyzing the breakdown of the levels of the Host factor within the levels of the Fungicides factor, a difference was observed only for the treatments DIMETOMORPH + MANCOZEB, ACIBENZOLAR-S-METHYL and MANCOZEB. Despite the difference observed, the percentage of emergence was higher than 91% for all Fungicide x Host combinations. Similarly, the absence of significant reductions in the percentage of emergence of the parasitoid's offspring has also been reported by other authors (CARMO et al. 2010; CARVALHO et al. 2012; PRATISSOLI et al. 2010b), indicating that this variable suffers little or no influence from fungicides.

4.4 CONCLUSION

The selectivity of the insecticides tested differed depending on the host under study, while the fungicides did not;

Selectivity results are influenced by the host, which implies the need to use target specimens when carrying out these tests.

4.5 REFERENCES

ALMEIDA, G. D.; ZANUNCIO, J. C.; PRATISSOLI, D.; ANDRADE, G. S.; CECON, P. R.; SERRÂO, J. E. Effect of azadirachtin on the control of *Anticarsia gemmatalisand* its impact on *Trichogramma pretiosum*. **Phytoparasitica**, v. 38, p. 413-419, 2010.

ALVES, F.R.; JESUS JUNIOR, W.C.; PRATISSOLI, D.; POLANCZYK, R.A.; ZANUNCIO JUNIOR, J.S.; HOLTZ, A.M.; VIANNA, U.R. Phytosanitary management of diseases and pests: new perspectives. In: JESUS JUNIOR, W.C.; POLANCZYK, R.A.; PRATISSOLI, D.; PEZZOPANE, J.E.M.; SANTIAGO, T. (Eds.). **Updates on plant health defense**. Alegre, Federal University of Espirito Santo, ch. 13, p. 383-416, 2007.

BESTETE, L.R; PRATISSOLI, D.; QUEIROZ, V.T. de; CELESTINO, F.N; MACHADO, L.C Toxicity of castor oil to *Helicoverpa zea* and *Trichogramma pretiosum*. **Pesquisa Agropecuaria Brasileira**, v.46, n.8, p.791-797, 2011.

CARMO, E.L. do; BUENO, A.F.; BUENO, R.C.O.F.; VIEIRA, S.S.; GOULART, M.M.P.; CARNEIRO, T.R. Selectivity of phytosanitary products used in soybean cultivation for pupae *of Trichogramma pretiosum* Riley, 1879 (Hymenoptera: Trichogrammatidae). **Arquivos do Instituto Biológico**, v. 77, n. 2, p. 283-290, 2010.

CARVALHO, G. A.; GODOY, M. S.; PARREIRA, D. S.; REZENDE, D. T. Effect of chemical insecticides used in tomato crops on immature *Trichogramma pretiosum* (Hymenoptera: Trichogrammatidae). **Revista Colombiana de Entomologia** 36 (1): 10-15 (2010).

CARVALHO, G. A.; PARRA, J. R. P.; BAPTISTA, G. C. Bioactivity of phytosanitary products used in tomato cultivation (*Lycopersicon esculentum* MILL.) to *Trichogramma pretiosum* Riley, 1879 (Hymenoptera: Trichogrammatidae) in F1 and F2 generations. **Ciência e Agrotecnologia**, v. 27, n. 2, p. 261-270, 2003.

CARVALHO, J. R.; PRATISSOLI, D.; PAES, J. P. P.; ZUIM, V.; STINGUEL, P. SELECTIVITY OF FUNGICIDES USED IN THE CULTURE OF TOMATOES (*Lycopersicum esculentum*, MILL.) TO *Trichogramma pretiosum*. **Nucleus** (Ituverava. Online), v. 9, p. 1-8, 2012.

CZEPAK, C.; FERNANDES, P.M.; ALBERNAZ, K.C.; RODRIGUES, O.D.; SILVA, L.M; SILVA, E.A.; TAKATSUKA, F.T.; BORGES, J.D. Selectivity of insecticides to the complex of natural enemies in the crop of cotton (Gossypium hirsutum). **Pesquisa Agropecuària Tropical**, v.35, n.2, p.123-127, 2005.

FERREIRA, D.F. **SISVAR**: version 5.1. Lavras: UFLA, 2003.

GIOLO, F. P.; GRÜTZMACHER, A. D.; MANZONI, C. G.; LIMA, C. A. B. NORNBERG, S. D. Toxicity of phytosanitary products used in peach cultivation on adults of *Trichogramma pretiosum*. **Bragantia**, v. 66, p.423-431, 2007.

GOULART, R.M.; BORTOLI, S.A.; THULER, de, R.T.; D. PRATISSOLI; VIANA, C.L.T.P.; H.X.L. VOLPE. Evaluation of the selectivity of insecticides to *Trichogramma* spp. (Hymenoptera: Trichogrammatidae) in different hosts. **Arquivos Instituto Biológico**, v.75, n.1, p.69-77, jan./mar., 2008.

HASSAN, S.A.; HAFES, H.; DEGRANDE, P.E.; HERAI, H. The side effects of pesticides on the egg parasitoid *Trichogramma caecoeciae* Marchal (Hym., Trichogrammatidae), acute dose-response and persistence tests. **Journal of Applied Entomology**, v. 122, n. 9-10, p. 569-573, 1998.

HASSAN, S.A.; HALSALL, N.; GRAY, A.P.; KUEHNER, C.; MOLL, M.; BAKKER, F.M.; ROEMBKE, J., YOUSEF, A.; NASR, F.; ABDELGADER, H. A laboratory method to evaluate the side effects of plant protection products on *Trichogramma cacoeciae* Marchal (Hym., Trichogrammatidae). In: CANDOLFI, M.P.; BLÜMEL, S.; FORSTER, R.; BAKKER, F.M.; GRIMM, C.; HASSAN, S.A.; HEIMBACH, U.; MEAD- BRIGGS, M.A.; REBER, B.; SCHMUCK, R.; VOGT, H. (Eds.). **Guidelines to evaluate side-effects of plant protection products to non-target arthropods.** IOBC/ WPRS, Gent. p. 107-119, 2000.

LUZ, J. M. Q.; SHINZATO, A. V.; SILVA, M. A. D. Comparison of conventional and organic tomato production systems in protected cultivation. **Bioscience Journal**, v. 23, p. 7-15,

2007.

MOURA, A. P.; CARVALHO, RIGITANO, G. A.; R. L. de O. Toxicity of insecticides used in tomato cultivation to *Trichogramma pretiosum.* **Pesquisa Agropecuaria Brasileira**, v.40, n.3, p.203-210, mar. 2005.

PRATISSOLI, D.; DALVI, L. P.; POLANCZYK, R. A.; ANDRADE, G. S.; HOLTZ, A. M.; NICOLINE, H. O. Biological characteristics of *Trichogramma exiguum* on eggs of *Anagasta kuehniella* and *Sitotroga cerealella*. **Idésia**, v. 28, n. 1, p. 39-42, 2010a. PRATISSOLI, D.; FORNAZIER, M.J.; HOLTZ, A.M.; GONÇALVES, J.R.; CHIORAMITAL, A.B.; ZAGO, H. Occurrence of *Trichogramma pretiosum* in commercial tomato areas in Espirito Santo, in regions of different altitudes. **Horticultura Brasileira,** Brasilia, v. 21, n. 1, p. 73-76, 2002.

PRATISSOLI, D.; MILANEZ, A. M.; BARBOSA, W. F.; CELESTINO, F. N.; ANDRADE, G. S.; POLANCZYK, R. A. SIDE EFFECTS OF FUNGICIDES USED IN CUCURBITACEOUS CROP ON *Trichogramma atopovirilia* OATMAN & PLATNER (HYMENOPTERA: TRICHOGRAMATIDAE). **Chilean Journal of Agricultural Research**, v. 70, n. 2, p. 323-327, 2010b.

PRATISSOLI, D.; MILANEZ, A. M.; CELESTINO, F. N.; BARBOSA, W. F.; VIANNA, U. R.; POLANCZYK, R. A.; ZINGER, F. D.; CARVALHO J. R. de. Selectivity of insecticides recommended for cucurbits for *Trichogramma atopovirilia* Oatman & Platner (Hymenoptera: Trichogrammatidae) under laboratory conditions. **Revista Ceres**, v. 58, n.5, p. 661-664, Sep/Oct, 2011.

PRATISSOLI, D.; VIANNA, U. R.; FURTADO, G. O.; ZANUNCIO, J. C.; POLANCZYK, R. A.; BARBOSA, W. F.; CARVALHO, J. R. Selectivity of insecticides to *Trichogramma pretiosum* Riley (Hymenoptera: Trichogrammatidae) in different hosts. **Boletin de Sanidad Vegetal Plagas**, v. 35, n. 3, p. 347-353, 2009.

RUSSELL, J. E.; STOUTHAMER, R. Sex Ratio Modulators of Egg Parasitoids. In: CÔNSOLI, F. L.; PARRA, J. R. P.; ZUCCHI, R. (Eds.) **Egg Parasitoids in Agroecosystems with Emphasis on *Trichogramma***. Springer, Piracicaba, p. 167190, 2010.

SCHOLZ B.C.G., ZALUCKI M.P. 2000. The effects of two new insecticides on the survival of adult Trichogramma pretiosum Riley in sweet corn. In: Austin A, Dowton M, editors. **Hymenoptera: Evolution, Biodiversity and Biological Control.** Canberra, Australia: CSIRO Pubs. p. 381-387.

SOUZA, J.C.; REIS, P.R. Main pests of table tomatoes: bioecology, damage and control.

Informe Agropecuàrio, v.24, p.79-92, 2003.

TROUGAKOS, I.; MARGARITIS, L. 2002. Novel morphological and physiological aspects of insect eggs. p. 3-36. In: HILKER M.; MEINERS, T. (Eds.). **Chemoecology of insect eggs and egg deposition**. Blackwell Publishing Ltd., Berlin, Germany.

VINSON, S. B. 1997. Host selection behavior of egg parasitoids, with emphasis on the Trichogrammatidae family, 67-120. In: PARRA, J. R. P., ZUCCHI, R. A. (eds.). ***Trichogramma* and applied biological control**. Piracicaba: FEALQ, 324pp.

WAJNBERG, E. Genetics of the Behavioral Ecology of Egg Parasitoids. In: CÔNSOLI, F. L.; PARRA, J. R. P.; ZUCCHI, R. (Eds.) **Egg Parasitoids in Agroecosystems with Emphasis on *Trichogramma***. Springer, Piracicaba, p. 149166, 2010.

ZANUNCIO, T.V., SERRÂO, J. E.; ZANUNCIO, J. C.; GUEDES, R. N. C. Permethrin-induced hormesis on the predator *Supputius cincticeps* (Stal, 1860) (Heteroptera: Pentatomidae). **Crop Protection**. v. 22, p. 941-947, 2003.

MIX
Papier aus verantwortungsvollen Quellen
Paper from responsible sources
FSC® C105338

Printed by Books on Demand GmbH, Norderstedt / Germany